Hazard and risk analysis in food processing

New approaches towards HACCP and food safety

Paul Besseling
Chitra Kalyankar
Elizabeth Montes Saavedra

Table of contents

Chapter 1: Standards for food safety management 4
1.1 Introduction 4
1.2 General Principles of Food Hygiene 5
1.3 European legislation 7
1.3.1 General principles of European food law 7
1.3.2 Legislation for food safety in companies 8
1.3.3 Regulations and directives regarding specific hazards 8
1.4 Certification standards for food safety 10
1.4.1 BRC Global Standard for Food Safety 10
1.4.2 IFS International Featured Standard 11
1.4.3 Safe Quality Food 11
1.4.4 ISO 22000 and FSSC 22000 11
1.4.5 GMP+ standard for animal feed 12
1.4.6 GlobalGAP 12
1.5 Comparison between standards 12

Chapter 2: Different approaches for risk analysis 16
2.1 Introduction 16
2.2 Risk assessment 17
2.3 Failure mode and effect analysis 18
2.4 The Bow Tie principle 19
2.5 Risk matrices 21

Chapter 3: Preparation of hazard and risk analysis 24
3.1 Introduction 24
3.2 The HACCP team 24
3.3 Prerequisite programmes 25
3.4 Description of products 26
3.5 Flow diagrams 26

Chapter 4: The principles of hazard and risk analysis. 28
4.1 Introduction 28
4.2 List of relevant hazard 29
4.3 Control measures 32
4.3.1 Prevention of hazards 32
4.3.2 Elimination or reduction of hazards 33

4.3.3 Food Safety Bow Tie 33

4.4 Critical control points 34

4.4.1 Deviations in control measures in the Bow Tie approach 35

4.5 Operational prerequisites programmes 36

4.6 Determination and validation of critical limits 40

4.7 Monitoring and corrective action 42

4.8 Validation and verification 44

4.8.1 Validation of a food safety management system 44

4.8.2 Verification of a food safety management system 45

4.9 Documentation and registration 47

4.10 Food Safety Bow Tie 48

Chapter 5: Risk matrix and decision tree. 50

5.1 Risk matrix 50

5.1.1 Severity of hazards 51

5.1.2 Probability of hazards 57

5.2 The use of a decision tree 61

5.2.1 CCPs and OPRPs in ISO 22000 62

5.2.2 Decision tree based on ISO 22000. 66

Chapter 6: Hazard and risk analysis in the food supply chain 70

6.1 Control in the food supply chain 70

6.2 Hazard and risk analysis of processes 71

6.3 Hazard and risk analysis of raw materials 73

6.4 Hazard and risk analysis for the use of product 75

Annex 1 - RECOMMENDED INTERNATIONAL CODE OF PRACTICE GENERAL PRINCIPLES OF FOOD HYGIENE CAC/RCP 1-1969, Rev. 4-2003 (1) 79

Annex 2 - HAZARD ANALYSIS AND CRITICAL CONTROL POINT (HACCP) SYSTEM AND GUIDELINES FOR ITS APPLICATION 93

Chapter 1: Standards for food safety management

The first chapter of this book focuses on the framework in which a hazard and risk analysis is performed. Section 1.1 discusses the developments that led to the current approach to food safety management. Section 1.2 covers the *General Principles of Food Hygiene*, the worldwide recognized code of practice for food hygiene and development of a HACCP based food safety management system. Section 1.3 and 1.4 deal with the relationship between legislation and certification standards for hazard and risk analysis. Section 1.5 compares different food safety standards with each other.

In this book, the term 'HACCP system' refers to the method which is described in the *General Principles of Food Hygiene* of the Codex Alimentarius. The term 'food safety management system' is referred to the system that is designed and implemented by food companies. The term 'hazard and risk analysis' refers to the analysis performed by food companies using HACCP system.

1.1 Introduction

The introduction of the HACCP system in the second half of the last century marks the beginning of a new chapter in the control of the safety of food. The principle of risk analysis along with the HACCP system is added to classical approaches such as 'end product testing' and 'processing in accordance with regulations'. History has extensively shown the limitations of these practices.

End product testing, which usually consists of sampling and microbiological or chemical analysis, can provide a useful tool to identify relatively frequent structural shortcomings (for example, 30 non-compliant products in a batch of 100). For detecting accidental errors (1 non-compliant product in a batch of 1000), the probability of detection by sampling is significantly lower. In order to increase the probability of detection, a large number of samples must be examined. However, from a financial perspective, this is not an appealing option. In addition to the representativeness of the samples, the sensitivity of the analysis techniques is also a common problem in end product testing. A practical problem is that some analysis are very time consuming, especially in relation to the short shelf life of some food. Moreover, end product testing is used to determine retrospectively whether something has gone wrong, while experience has underpinned the need for a preventive approach rather than reactive.

The second classical approach, 'processing in accordance with regulations' has a long history followed by norms, directives, statutes and codes. In fact, in this approach, a group of experts, commissioned by authorities, establishes the rules, methods and conditions under which food shall be produced. This led to the development of standards for different types of products such as meat, dairy, canned food, confectionery, etc. However, this approach failed to withstand the stormy and continuous development of the food supply chain in the twentieth century. The emergence of new packaging and storage technologies, new preparation techniques and food trade have ensured an expansive food market growth.

Drafting legislation faces the challenge of developing requirements for an ever-growing number of newly created products. The development of laws or amendments to laws requires specific knowledge and deep insight of products and processes, which sometimes may be lacking; general regulations may not provide sufficient control of specific risks. Moreover, policy makers that take the responsibility for drafting regulations and conducting subsequent inspections, also take a responsibility for the safety of the product. In the approach of today's society however, the safety of products is primarily a responsibility of the producers and not for external authorities. In short, in the course of the twentieth century, the need arose to approach the problem of unsafe food in a different way.

The transition with the traditional approach is strikingly illustrated in the fourth edition of the '*General Principles of Food Hygiene*' published in 2003 as part of the Codex Alimentarius. The body of the document consists of a set of general requirements for producing safe and sound food and thus, fits into the conventional approach. On the other hand, the annexed document titled, '*Hazard Analysis and Critical Control Point (HACCP) system and guidelines for its application*' introduces the modern approach of risk analysis.

This book describes the requirements for hazard and risk analysis from different laws and standards. These requirements are combined with general principles to perform risk analysis in other types of industry; this will ultimately result in a practical approach to conduct hazard and risk analysis suitable for the food industry. A brief description and comparison of food regulations are discussed in this chapter. Different approaches of risk analysis are described in chapter 2. Chapter 3 contains the activities, which must be performed in the preparation of hazard and risk analysis. In chapter 4, the principles and definitions of the HACCP system are coupled with ISO 22000 and a method of risk analysis, which is known as the Bow Tie principle. In Chapter 5, the use of risk matrices and decision trees is discussed. Chapter 6 deals with the registration and documentation of hazard and risk analysis, and risk management towards raw materials and suppliers and towards buyers of the products.

The text from the *General Principles of Food Hygiene* is used as the basic document to describe the principles of HACCP. This text is included as appendix 1 of this book.

1.2 General Principles of Food Hygiene

For food companies, the *General Principles of Food Hygiene* is the most important document regarding the safety and soundness of food products. The document is part of a collection of international standards for food production that have been drafted by the Codex Alimentarius Commission. This document, together with the *'Principles for the establishment and application of microbiological criteria for foods'* and the *'Principles and guidelines for the conduct of microbiological risk assessment'* forms the basis of international laws and standards for food safety.

The Codex Alimentarius Commission (CAC) was established in 1962 as an initiative of the United Nations and compiled from the World Health Organisation (WHO) and the Food and Agricultural Organisation (FAO). CAC was initiated in response to the ever-growing international trade, thereafter resulting in the need for standards and legislation to harmonize food safety control. The main task of CAC is to develop and update regional and global standards and guidelines; among others, its scope encompasses safety and hygiene, processing, storage, labelling, quality and packaging of food. The overall aim of the Codex Alimentarius is to ensure that different countries have a common basis for their legislation regarding food production and food trade. The European Union also makes extensive use of the standard in the Codex Alimentarius. For example, the legislation for food labelling is based on the Codex document *'General Standard for the labelling of pre-packaged Foods'*. The establishment of the European Food Safety Authority (EFSA), which is governed by European Regulation (EC) 178/2002, is also based on the scientific insight of the document *'Principles and Guidelines in the conduct of microbiological risk assessment'*.

The first version of the *'General Principles of Food Hygiene'* was published in 1969. In 1993, the principles of the HACCP system were incorporated as an annex. The current fourth version is published in 2003. The *'General Principles of Food Hygiene'* forms the basis of European Regulation (EC) 852/2004 on the hygiene of foodstuffs and European Regulation (EC) 183/2005 laying down requirements for feed hygiene.

The contents of the General Principles of Food Hygiene are as follows:

1. Objectives
2. Scope, use and definition
3. Primary production
4. Establishment: design and facilities
5. Control of operation

6. Establishment: Maintenance and Sanitation

7. Establishment: Personal Hygiene

8. Transportation

9. Product information and consumer awareness

10. Training

Annex: Hazard Analysis and Critical Control Point (HACCP) System and guidelines for its application.

Chapter 1 and 2 are introductory chapters. Chapters 3 to 10 are referred to as 'prerequisite programs'. They contain general requirements regarding, for example, maintenance and cleaning of buildings and equipment, personal hygiene and training of employees.

In Chapter 1, the objectives of the *General Principles of Food Hygiene* are described as:

1. Identify key principles of food hygiene applicable throughout the food chain (including primary production to final consumer) in order to ensure that food is safe and suitable for human consumption.

2. Recommend a HACCP based approach as a mean to strengthen food safety.

3. Indicate how the principles of food hygiene can be put into practice.

4. Provide guideline to develop specific requirements. This may be required for different sectors along the food chain, for example, production, processing or control of raw materials and/or to strengthen existing hygiene specifications.

Chapter 2 indicates how the *'General Principles of Food Hygiene'* can be used in practice, and describes the relationship between prerequisite programs and the HACCP system. The chapter clarifies that Chapters 3 to 10 contain *'general'* principles and thereby suggest: *'There will inevitable be situations where the specific requirements in this document are not applicable. The fundamental question in every case is 'what is necessary and appropriate on the grounds of the safety and suitability of food for consumption?'*

The text indicates where such questions are likely to arise by using the phrases "where necessary" and "where appropriate". In practice, this means that although requirements are generally applicable and reasonable, there will be situations where the requirements will be neither necessary nor suitable. In order to determine whether a requirement is necessary or suitable, risk assessment, preferably within the framework of HACCP approach, should be carried out. This approach enables the application of requirements in a flexible and sensible manner; in compliance with the ultimate goal of producing safe food apt for consumption. In this way, a wide range of activities and different levels of risks in food production are taken into account. The 'General Principles of Food Hygiene' are thus considered as a basis for additional guidelines or standards for specific products or sectors in the food chain. Also, see the fourth objective of the 'General Principles of Food Hygiene'.

Example
'Where necessary' and 'where applicable'
General Principles of Food Hygiene paragraph 4.2.2: Internal structures and fittings. Windows should be easy to clean, be constructed to minimize the buildup of dirt and where necessary, be fitted with removable and cleanable insect-proof screens. Where necessary, windows should be fixed.
End of example

A large number of specific prerequisite programs are included in Codex Alimentarius. For example, *Code of Hygienic Practice for Foods for Infants and Children, Code of Hygienic Practice for Groundnuts (Peanuts), General Principles of Meat Hygiene* and *Code of Hygienic Practice for the Collection, Processing and Marketing of Natural Mineral Waters*. The prerequisite program from the *General Principles of Food Hygiene* has served as a basis for Annex II of European Regulation (EC) 852/2004. It is included as literal translation in the Requirements for a HACCP based food safety system. Moreover, it forms a model for standard like the BRC standard and ISO 22002-1 (see paragraph 1.4.5).

The annex *Hazard analysis and Critical Control Point (HACCP) System and guidelines for its application* consists of two sections: the first section includes introduction, definitions and principles. The second section is a guideline for application of the HACCP system.

1.3 European legislation

The European Union has published various regulations and directives, which have a close relationship with the hazard and risk analysis for food safety. Hereof Regulation (EC) 852/2004 is the most important. The following paragraph depicts different regulations and guidelines and their relationship with HACCP.

1.3.1 General principles of European food law

The general principles and requirements of European food law are laid down in European Regulation (EC) 178/2002. The regulation includes provisions regarding the European Food Safety Authority (EFSA), the rapid alert system for food and feed (RASFF), crisis management and emergency situations. The regulation is also known as the *General Food Law*. The regulation specifies that European food law aims to provide food policies that can assure:

- a high level of protection of human life and health

- the protection of consumers interests, including fair practices in food trade

- the achievement of the free movement in the European Community of food and feed manufactured or marketed according to the general principles and requirements.

In order to pursue a high level of protection of human life and health, food law shall be based on the results of risk analysis. Risk analysis incorporates risk assessment, risk management and risk communication. Risk assessment is a process that assesses risks in an independent, objective and transparent manner based on scientific data. The principles of risk assessment in European Regulation (EC) 178/2002 are partly based on the '*Principles and guidelines for the conduct of microbiological risk assessment*' of the Codex Alimentarius. Risk management is the process of weighing policy alternatives with regard to the results of risk assessment and, if required, selecting and implementing appropriate control options, including regulatory measures, if necessary, in consultation with stakeholders.

Risk communication is an interactive exchange of information and opinions concerning hazards and risk among risk assessors, risk managers, consumers, producers, academia, and other stakeholders. The EFSA was established to bring risk assessment into practice. The significance of risk assessment is further discussed in section 2.2. The EFSA provides independent and scientific advice and technical support to the legislation and policies within the European Community regarding the safety of food and feed.

The definitions and provisions regarding risk analysis in Regulation (EC) 178/2002 are focused on the activities of the legislature and not that of food companies. Regulation (EC) 178/2002 is thus not directly relevant to the application of HACCP in food companies. Section 2.2 of this book discusses the relationship between risk assessment and risk analysis in the HACCP system. Regulation (EC) 178/2002 incorporates definitions and conditions regarding unsafe food. For example: traceability in article 18 and withdrawal of unsafe products from market in article 19.

Chapter 4 of Regulation (EC) 178/2002 describes the rapid alert system, crisis management and emergencies. Information from the *Rapid Alert System for Food and Feed (RASFF)* is published on the website of European Union. In the context of HACCP, food companies can use the information from RASFF to identify hazards. See also section 4.2 List of relevant hazards.

1.3.2 Legislation for food safety in companies

With the entry into force of Council Directive 93/43/EEG of 14 June 1993 on the subject of food hygiene, application of HACCP has become a general legal obligation in the European Union starting December 15, 1995. In April 2004, European Regulation (EC) 852/2004 replaced this directive. Regulation (EC) 852/2004 consists of the regulations for food companies concerning hygiene and HACCP system. Article 4 'General and Specific Hygiene Requirements' indicates the obligation for compliance with the general hygiene specifications listed in Annex II of the Regulation.

This annex includes requirements concerning company premises, transport, food waste, personal hygiene, packaging and training. It is partly based on *General Principles of Food Hygiene* of the Codex Alimentarius. The principles of HACCP in article 5 'Hazard analysis and critical control points' are also adopted from the *General Principles of Food Hygiene*. Regulation (EC) 852/2002 hereby gives no additions or changes. Both the *General Principles of Food Hygiene* and Regulation (EC) 852/2004 recognize the indications from governments and industries regarding barriers in the effective implementation of the HACCP by individual companies. This is especially true in case of small and less developed companies. Hence, in Regulation (EC) 852/2004, the application of the HACCP is facilitated by offering the companies a possibility of using 'guides for good practice'. These guides can be created in collaboration with different companies, for example, through trade organizations or through the marketing boards. The guides should be made in consultation with representatives of stakeholder parties, such as food safety authorities, consumer associations and business associations. Regulations regarding preparation, distribution and use of national and community guides are depicted in articles 7, 8 and 9 of Regulation (EC) 852/2004. In the European Union member states, approximately 500 guides for good practice were recognized up to 2009.

European Regulation (EC) 853/2004 contains specific hygiene specifications for food of animal origin. This regulation complements the requirements in Regulation (EC) 852/2005. The regulation consists of twelve sections for different animal products such as meat of farm animals, poultry and game, fishery products, raw milk and dairy products, eggs and egg products, and animal products such as gelatin and collagen.

European Regulation (EC) 183/2005 lays down requirements for feed hygiene. Concerning hygiene and HACCP, the scope is similar to that of Regulation (EC) 852/2004. Regulation (EC) 183/2005 also incorporates some specific provisions such as recognition of feed companies, which use, manufacture or sell additives.

1.3.3 Regulations and directives regarding specific hazards

The European legislation includes a number of regulations and directives that relate to specific hazards. This section of this book successively discusses regulations and directives regarding microbiological, chemical and physical hazards, and finally allergens.

Rules and criteria for microbiological hazards are included in European Regulation (EC) 2073/2005. Wherever applicable this regulation shall be an integrated part of hazard and risk analysis of the food safety system. The microbiological criteria in the regulation shall be used in the validation and verification of individual control measures. The regulation provides food safety criteria and process hygiene criteria. The food safety criteria comprises of standards for different organisms - and potential toxins - in different products during the entire shelf life. The microbiological standards in the process hygiene criteria relate to certain stages in processing.

If microbiological research led to an inadequate result regarding food safety criteria, the food company must prevent the release or ensure the withdrawal of the product form market. Furthermore, measures should be taken to detect the cause of the inadequate results in order to prevent recurrence. Inadequate results concerning process hygiene criteria need not directly commence the blockade of product provided further research demonstrates that the final products do comply with the food safety criteria. Measures should be undertaken to ensure that a subsequent production will adhere to the process hygiene criteria. Examples of measures are improving the production hygiene, reviewing the process management, improvement in the selection of raw materials, monitoring the efficiency of heat treatment, and prevention of recontamination.

The food safety criteria comprises of standards for hazards such as *Listeria monocytogenes* and *Salmonella* for products like infant formula, ready-to-eat food, food for special medical purposes, minced meat and meat products, dairy products based on unpasteurized milk, egg products, cooked crustaceans (seafood), sprouted seeds (ready-to-eat), pre-cut ready-to-eat vegetables and fruits and unpasteurized fruits and vegetable juices. The requirements concerning presence of *Staphylococcus*-enterotoxins are indicated on cheese, milk powder and whey powder. The requirements regarding *Cronobacter sakazakii* are indicated on dry infant formula and dry dietary food for special medical purposes. The requirements regarding presence of *histamine* apply for fishery products. In a number of cases, for process hygiene criteria, the requirements concern more general microbiological values such as total aerobic-bacterial colony count and presence of indicator organisms as *Enterobacteriaceae* for meat and meat-products, milk and dairy products and egg-products. Process hygiene criteria are established for meat and meat products, milk and dairy products and egg products, fishery products, vegetable and fruits and thereof derived products.

European Regulation (EC) 1881/2006 is predominantly important concerning chemical hazards. The regulation comprises of standards for the maximum levels of different hazards in various products. For example, standards for nitrate in leafy vegetables such as spinach and lettuce, for mycotoxins such as aflatoxin and ochratoxin in grains and nuts, for metals such as inorganic tin from tin packaging, lead, cadmium and mercury, for dioxins and for poly-aromatic hydrocarbons (PAHs).

Food, in which contaminants exceed the maximum level, shall not be released for the market. They also should not be used as an ingredient and they should not be mixed with batches that comply with the maximum levels. However, in some cases, it is permissible to reduce the level of contaminants. For example, by using optical separation methods, the level of aflatoxins in grains can be reduced and PAHs are extracted from edible oil by the use of activated carbon. Food contaminated with mycotoxins, may not be intentionally treated with chemicals for removing these impurities. In relation to hazard and risk analysis, Regulation (EC) 1881/2006 establishes the standards for the validation of control measures.

European Regulation (EC) 1935/2004 is applicable for chemical contaminants from packaging material and other materials intended to come into contact with food. The regulation includes a positive list of materials, which are legally permitted as packaging materials for food. The general principle here is that the interaction between contact material and food is limited so that migration limits are not exceeded. Specific European legislation is applicable to some of these materials. For example, plastics in Regulation 10/2011, vinyl chloride in PVC in Directive 78/142/EC, epoxy coating in Directive (EC) 1895/2005 and inorganic tin in can packaging in Directive (EC) 1881/2006.

Moreover, the European Union member states have national legislation for individual materials. Food producers by means of a declaration of compliance or specifications and/or statements must declare that the used materials are suitable for packaging or as contact materials and that legal requirements are met.

For physical hazards, there is no specific European legislation. It is unfortunate that consumers often detect physical contaminants such as pieces of glass, metal, rubber or wood in their food. This requires constant attention of food companies. Generally speaking, food should be free from foreign materials or objects. The only European provision that addresses foreign bodies is article 14 of Regulation (EC) 178/2002 – food safety requirements. This article states that: '*In determining whether any food is unfit for human consumption, regard shall be had to whether the food is unacceptable for human consumption according to its intended use, for reasons of contamination, whether by extraneous matter or otherwise, or through putrefaction, deterioration or decay.*'

The fourth and the last category of hazards comprises of allergens. European legislation regarding allergens is a part of Regulation (EC) 1169/2011, which includes regulations about labelling, and presentation of food. The regulation lists the allergens for which an indication on the label of consumer product is compulsory, whenever these allergens are naturally present in one of the ingredients. Food producers, by controlling recipes and specifications must make it evident that the allergens are correctly indicated on all labels of the products. Hazard and risk analysis shall ascertain appropriate measures are implemented to prevent contamination with allergens foreign to the product.

1.4 Certification standards for food safety

Following the introduction of the legal obligations to apply HACCP in 1995, a need arose in food companies to have their self-developed food safety system assessed by independent experts. From this, an idea arose to create a certification standard such as ISO 9000, which was already developed for quality management systems. In the Netherlands, in 1996, this led to the publication of 'Assessment criteria for an operational HACCP system', the first version of the later 'Requirements for a HACCP-based Food Safety System'. Internationally this standard is referred to as "Dutch HACCP" (*Eisen voor een op HACCP gebaseerd voedselveiligheidssysteem*). Similar initiatives started in other countries. In 1994, Safe Quality Food (SQF) was published in Australia. The Danish standard DS 3027, which has served as a model for the development of ISO 22000, was published in 2005. The BRC standard (BRC: British Retail Consortium) was developed in 1998. In this case, the initiative came from retailers who, as part of their product liability, felt obliged to be more active in assessing their suppliers. In 2002, German and French retailers published the IFS-standard (IFS: International Featured Standard). The development of the BRC and IFS standard is also driven by the growing importance of private-label-products, which rendered retail organizations to become more sensitive about damage to their reputation.

An important initiative regarding certification standards was taken in 2002 with the establishment of Global Food Safety Initiative (GFSI). The initiative for GFSI came from CIES (Comité International d'Entreprises à Succursales), an association of retail organizations founded in 1954. The aim of GFSI is to establish an accreditation scheme for certification standards in order to ensure that all retailers will accept suppliers, who are certified based on one of the GFSI recognized standards. 'Once certified, accepted everywhere' is the slogan. This is to prevent food producers to be confronted with different certification requirements from different retailers. The three core elements of GFSI-accreditation scheme are the pre-requisite programs, hazard and risk analysis in accordance with HACCP and the principles for quality management based on ISO 9000. A list of GFSI recognized certification schemes could be found on the GFSI-website. In practice, the aim, 'once certified, accepted everywhere' has a limited significance. Several retailers unabatedly continue to develop their own standards and create additional requirements.

The sequel of this section depicts a number of leading certification standards and their relation to hazard and risk analysis in more detail.

1.4.1 BRC Global Standard for Food Safety

In the 90s a number of retailers established the British Retail Consortium to strengthen their position in relation to product quality. They developed the BRC Global Standard for Food Safety and subsequently, demanded that food suppliers should meet this standard. The first version of BRC standard was published in 1998; the seventh was published in 2015.

The BRC Standard includes requirements related to quality management, application of HACCP and implementation of prerequisite programs. In the BRC standard, the prerequisite programs are translated into four separate chapters with requirements regarding site, product, process and personnel.

As compared to Dutch HACCP, the BRC standard is less profound with regard to the principles of HACCP and most of the requirements relate to prerequisite programs. With almost 9000 certificates in 2007, the BRC standard is one of the most widely used standard in the world. In 2001 BRC has developed a *Global standard for packaging and packaging materials* for (food) packaging industry; this is called BRC IoP or BRC-packaging. In 2006, the *BRC Global Standard Storage and Distribution* was published for storage and distribution of food. In 2014, the BRC global standard for agents and brokers was published.

1.4.2 IFS International Featured Standard

Following British Retail Consortium, the first version of International Featured Standard (IFS) was published in 2002 by the cooperation of German and French retailers. In 2012, the sixth version was published. Italian retailers also joined in the fifth version.

Content wise the IFS standard resembles the BRC standard in many aspects. Alike BRC, IFS also requires a quality management system, application of HACCP and implementation of prerequisite programs. IFS differs from BRC particularly in the audit protocol and the way in which requirements are assessed.

The number of certificates (around 8500 in 2007) issued based on IFS standard was slightly lower than the number of BRC-certificates. Alike the BRC standard, the certificates are evenly distributed across the world. Since May 2006, it is possible to certify transport companies against IFS Logistics. In 2009, the first version of IFS Broker was published.

1.4.3 Safe Quality Food

Safe Quality Food (SQF) is an assessment standard developed in Australia. SQF is designed for all types of food companies and works with various standards for different phases in the chain:

- SQF 1000 for primary agricultural sector and small-scale processors;

- SQF 2000 for 'larger' processing industry;

- SQF 3000 specifically aimed at retail.

HACCP is one of the basic elements of the standard. Besides critical control points for food safety (CCPs), SQF also includes critical points for quality (CQPs). Like BRC and IFS, SQF has requirements for quality management, food safety and prerequisite programs.

The Food and Marketing Institute (FMI) in Washington manages SQF since the summer of 2003. In February 2009, during the CIES Food Safety Conference, SQF and IFS announced that they joined forces. They will cooperate in the field of selection of accreditation and certification bodies.

SQF has gained little foothold in Europe. More than 10,000 certificates are issued collectively in Asia, Europe, the Middle East, North and South America.

1.4.4 ISO 22000 and FSSC 22000

In recent years, several standards were developed for assessment and certification of food safety management systems. This is sadly witnessed by the international food businesses. The need arose to halt this growing number of standards and establish an internationally accepted standard. This led to the publication of ISO 22000 in September 2005. The standard is managed and published by the International Organization for Standardization (ISO). The secretariat of ISO is located in Geneva. About thirty countries have contributed in the development of ISO 22000 and hence it has enormous international support. The Codex Alimentarius Commission is also involved in the development of ISO 22000.

ISO 22000 follows the structure of ISO 9001 and integrates it with food safety systems based on HACCP principles. ISO 22000 does not intend to replace ISO 9001. ISO 22000 includes only food safety requirements, while ISO 9001 also includes quality aspects. Since ISO 22000 is a standard to assess management systems, the requirements are less specific than BRC and IFS, which are designed as inspection standards.

The standard can be applied by all organizations, which are directly or indirectly part of the food supply chain. For smaller and/or less developed businesses (such as small cheese farms or catering companies) ISO 22000 offers the possibility to introduce an externally developed combination of control measures. This approach has similarities to the concept of 'guides to good practice' in Regulation (EC) 852/2004. ISO 22000 can also be used by organizations that are indirectly involved in the food supply chain, such as suppliers of machinery and equipment, and suppliers of packaging materials.

To be eligible for the GFSI recognition, the FSSC 22000 (Food Safety System Certification) standard was published in October 2009. FSSC 22000 integrates the requirements of ISO 22000 and requirements in prerequisite programs in the ISO 22002-series. The ISO 22002 series includes prerequisite programs for food processing, catering, primary production and packaging materials.

1.4.5 GMP+ standard for animal feed

The GMP+ standard was prepared by the Dutch Central College of Experts for Feed Sector (Centraal College van Deskundigen Diervoedersector, CCvDD). The abbreviation GMP stands for Good Manufacturing Practice. After recent crisis (e.g. BSE), this feed standard was extended with the requirements for application of HACCP. The standard was then referred to as GMP+HACCP, later abbreviated to GMP+. The standard was originally intended for use in the Netherlands but gradually companies from more than 50 countries worldwide are certified according to GMP+.

The GMP+ standard is based on prerequisites programs in combination with requirements from ISO 9001, ISO 22000 and HACCP principles. A generic risk assessment is included in a database that contains information which can be used by companies for the purpose of hazard and risk analysis. The database is accessible to the GMP+ certified companies.

1.4.6 GlobalGAP

In 1997, a number of European retailers took the initiative to achieve harmonization of the requirements for suppliers of fresh products. This resulted in the Global Good Agricultural Practice Standard, commonly referred to as GlobalGAP. GlobalGAP lays down requirements for famers and horticulturists on food safety, sustainability and quality.

Food safety is the most important component of GlobalGAP. GlobalGAP includes requirements for animal welfare, environment and the wellbeing of employees. GlobalGAP includes a standard for feed producers. The ultimate goal is that GlobalGAP – as an international standard - becomes an umbrella standard for food safety and quality for primary production.

1.5 Comparison between standards

In management system, whether they are designed to manage food safety, quality, work safety or environment, three elements can be distinguished:

- Prerequisites or technical standards;
- Risk analysis;
- System management.

For food safety, the prerequisites include requirements related to buildings, equipment, machinery and tools, facilities for processes, facilities for employees, and so on. Prerequisites are incorporated in the *General Principles of Food Hygiene* or in Annex II of European Regulation (EC) 852/2004. Risk analysis consists of an analysis of hazards and risks and associated measures such as monitoring programs and handling nonconforming products. System management covers subjects arising from principles for quality management (see ISO 9000) such as management assessment, international audits, document management and training of employees.

The prerequisite programs - as in chapters 3 to 10 of the General Principles of Food Hygiene - contain approximately 150 different requirements. In table 1.1 (Requirements in the prerequisite programs), different requirements from these prerequisite programs are classified in three elements.

Table 1.1 Requirements in the prerequisite programs

Requirements related to the control of hazards: approximately 100 measures aimed at prevention of contamination and approximately 15 aimed at prevention of growth of microorganisms and /or elimination or reduction of hazards.	Approximately 115
Requirements related to risk analysis (including monitoring and corrective actions, recall and traceability).	Approximately 17
Requirements related to management (documentation, communication, training, supervision and leadership).	Approximately 17
Total requirements	Approximately 150

This analysis demonstrates that about 75% of the requirements are related to the prevention or reduction of hazards. They should be referred to as control measures as per the definition of HACCP. Control measures are further discussed in chapter 4.

Example
"Examples of prerequisites in the General Principles of food hygiene"

- Requirements to prevent contamination:

In 4.2.2: Structures within food establishments should be soundly built of durable materials and be easy to maintain, clean and where appropriate, able to be disinfected.
In 6.3.2: Buildings should be kept in good repair and condition to prevent pest access and to eliminate potential breeding sites. Holes, drains and other places where pests are likely to gain access should be kept sealed.

- Requirements to prevent development:

In 4.4.5: Depending on the nature of the food operations undertaken, adequate facilities should be available for heating, cooling, cooking, refrigerating and freezing food, for storing refrigerated or frozen foods, monitoring food temperatures, and when necessary, controlling ambient temperatures to ensure the safety and suitability of food.

- Requirements to eliminate or reduce hazards:

In 5.2.5: Physical and chemical contamination: in manufacturing and processing, suitable and effective detection or screening devices should be used where necessary.

- Requirements for risk analysis and control programs:

In 4.3.2: Equipment should be designed to allow temperatures to be monitored and controlled.
In 5.1: Food business operators should control food hazards through the use of systems such as HACCP.
In 6.5: Sanitation systems should be monitored for effectiveness, periodically verified by means such as audit pre-operational inspections or, where appropriate, microbiological sampling of environment and food contact surfaces and regularly reviewed and adapted to reflect changed circumstances.

- *Requirements for management of system:*

In 5.7: *Where necessary, appropriate records of processing, production and distribution should be kept and retained for a period that exceeds the shelf-life of the product.*
In 10.1: *All personnel should be aware of their role and responsibility in protecting food from contamination and deterioration.*

End of example

The result of analysis in table 1.1 is presented in figure 1.1 (Graphical representation of comparison between standards). Prerequisites according to the General Principle of food hygiene are placed in the right side top of the figure. Other standards in figure 1.1 are analysed in the same way. For example, of approximately 300 individual requirements form ISO 22000-standard, 27% are related to the management of the system, 63% to hazard and risk analysis and 10% to prerequisites. In the IFS standard, 35% of the requirements relate to prerequisites, 25% at hazard and risk analysis and 20% at the management system.

To some extent, the allocation of different requirements to different elements is subjective and debatable. Nonetheless, it gives an impression of the final image of comparison between the certification standards. Figure 1.1 displays the differences between the inspection standards like BRC and IFS and the system oriented standards like ISO 22000. FSSC 22000 is the most comprehensive (around 480 individual requirements) and perhaps also the most well balanced standard. The figure includes the third version of Dutch HACCP and displays the difference between BRC and IFS.

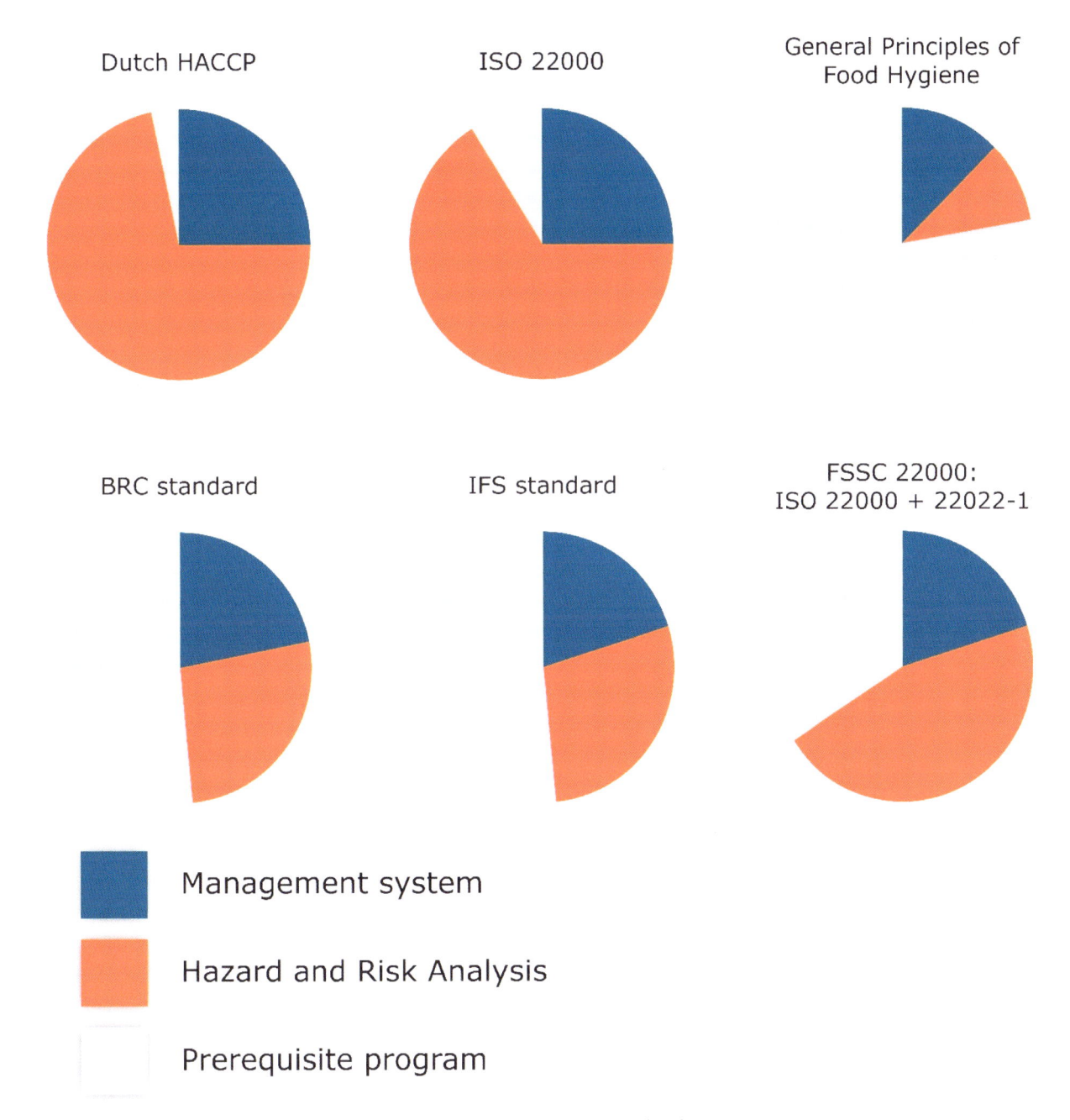

Figure 1.1: Graphical representation of comparison between standards

Chapter 2: Different approaches for risk analysis

This chapter throws light on different approaches for assessment of risks: Risk evaluation, Failure Mode and Effect Analysis (FMEA) and the Bow Tie Principle. Risk matrices are also discussed.

2.1 Introduction

The aim of risk analysis is to apply a structured approach to identify and assess risks and to undertake measures to reduce them. Assessment makes it possible to compare different risks to facilitate effective deployment of available resources for the implementation of control measures. In the process of risk analysis one generally seeks answers for the following questions:

- What can go wrong?

- What is the severity and probability of that?

- Which measures can be undertaken?

The core of risk analysis is the way in which the term 'risk' is defined. European Regulation (EC) 178/2002 gives the following definition:

Risk: A function of the probability of an adverse health effect and the severity of that effect, consequential to a hazard.

ISO 22000 states: 'the term "hazard" should not be confused with the term "risk" which, in the context of food safety means a function of the probability of an adverse health effect (become sick), and the severity of that effect (death, hospitalization, absence from work, etc.), when exposed to a specific hazard'. ISO 22000 hereby refers to ISO/IEC Guide 51:1999 Safety aspects – Guidelines for their inclusion in standards in which the term 'risk' is defined as 'combination of the probability of occurrence of harm and the severity of that harm'.

In this book, the term 'hazard' is used in the sense of an 'undesirable substance in food which can cause harm to the health of consumer'. The term 'risk' is used to indicate 'the severity of harm caused by a hazard and the probability of occurrence of that harm'.

The definition of risk can also be expressed in the following formula:

Risk = probability × severity

The simplicity of this formula is attractive yet deceptive. Behind the probability and severity, a number of situations, circumstances, causes, consequences, measures, vulnerabilities, probabilities and uncertainties are hidden. The numerous publications on the subject of risk analysis put forth two elements in varying degrees. They are:

1. The logic of causes and consequences. Risk analysis tries to identity the relationship between causes and effects in a structured manner.

2. Assessment of probability and severity. Risk analysis attempts to determine the magnitude of risk. The possibilities for this are a strict quantitative approach at one end and a complete qualitative approach at the other end.

A number of interesting approaches for risk assessment is discussed in the following paragraphs.

2.2 Risk assessment

Risk assessment is a scientifically based method to assess risks and its contributing factors. For microbiological hazards, the method is incorporated in the Codex Alimentarius under the title Microbiological Risk Assessment (MRA). The MRA-document is one of the basic documents regarding food safety. MRA intentionally focuses on the assessment of risks of microbiological hazards; however, the principles are also useful for assessment of other hazards. Conducting risk assessment as per MRA is primarily a governmental charter. The Codex document indicates thereon in its introduction: "*This document is primarily of interest to governments, although other organizations, companies and other interested parties who need to prepare microbiological risk assessment will find it valuable*".

Only governments and larger food enterprises will be able to gather the knowledge in teams made up of microbiologists, toxicologists, physicians and food technologists which can accomplish such a mission successfully. In Europe the establishment of EFSA (European Food Safety Authority), introduced in Regulation (EC) 178/2002 has directly resulted from risk assessment. Within EFSA different active commissions (scientific panels) are responsible for communicating scientifically justified opinions concerning safety of, for example, packaging materials, ingredients and genetically modified organisms.

Risk assessment is a component of 'risk analysis' as indicated in Regulation (EC) 178/2002. Next to risk assessment, risk analysis includes two other components namely risk management and risk communication. Risk assessment is a process consisting of the following four components:

- Hazard identification. A microorganism or toxin which forms a subject of risk assessment is determined and defined.

- Exposure assessment. Exposure is assessed on the basis of possible degree of contamination combined with detailed diet information. Determination of exposure makes use of food consumption surveys, residue levels, early contamination, process and distribution variables, and use by consumers.

- Hazard characterization. The aim of hazard characterization is to determine, in qualitative and/or quantitative sense, the severity and duration of adverse health effects associated with the hazard. Insights about this can be obtained by establishing a dose-response relationship.

- Risk characterization. In risk characterization the outcomes of exposure assessment and hazard characterization are combined and converted into qualitative and quantitative estimates concerning probability and severity of adverse health effects of the identified hazard. The reliability of risk characterization depends on uncertainties and estimates in the previous steps.

The conclusions of risk assessment are translated into standards and measures which lead to reduction and/or control of the hazard and/or risk. In risk analysis this is called risk management. The third component, risk communication, refers to the exchange of information and opinions regarding hazards and risks, among risk assessors, risk managers, consumers, food and feed businesses, academic world and other stakeholders.

Example of risk communication to consumers:
'Beware; give no chance to harmful bacteria. Thus ensure that these bacteria do not end up in your food via packaging, your hands or kitchen utensils. Cook this flesh thoroughly in order to eliminate these bacteria.'

This warning in mandatory in the Netherlands on all prepacked poultry meat.
End of example

In risk assessment the problem of unsafe food is approached from the perspective of a population or group of consumers. This contradicts the HACCP system in which a product and the corresponding process are considered as a starting point. This implies that risk assessment as such is not useful in the context of HACCP.

However, the outcome of risk assessment could be that a standard is set for the maximum level of a certain hazard and subsequently forming the aim of a HACCP based food safety system. Moreover, risk assessment generates information about exposure, health effects and dose-response relations. This information is important to determine the severity and probability of hazards, which is a part of HACCP.

Risk assessment is important when it comes to the ultimate objective of food safety management systems. Setting parameters for control measures in a food safety management system ideally begins with the description of a risk which can be tolerated from a safety perspective. For this purpose, the government determines an Appropriate Level of Protection (ALOP). An ALOP describes the number of illnesses (cases) that are considered as acceptable for a particular hazard. An example of an ALOP: government strives for a maximum number of two infections with Listeria monocytogenes per one million inhabitants per year. An ALOP can be subsequently translated into a *Food Safety Objective* (FSO or 'food safety criterion' as in Regulation (EC) 2073/2005). A FSO stands for the maximum frequency and/or concentration of a hazard in food at the moment of consumption. A FSO can be then translated into process criteria (PC). A process criterion indicates the acceptability of a product at a particular stage in a process or for a particular control measure, based on presence or absence of a hazard. Regulation (EC) 2073/2005 is a good example of how standards are set by risk assessment, which can be used during application of HACCP.

2.3 Failure mode and effect analysis

Failure mode and effect analysis (FMEA) is a method in which risks of potential failure are estimated by means of brain storming sessions. Accordingly, measures can be undertaken to control these risks. Development of this method was initiated by the American army in around 1949. Over the course of years, this method has been modified and improved. FMEA is an integral part of quality standard QS 9000, which is a joint venture of three American automobile manufacturers. FMEA is applied in many different industries such as automotive and aerospace. The characteristic feature of this method is the probability and effect diagram that serves as a tool to calculate and categorize a risk.

The core of FMEA method is based on the question: 'Which failure can occur and what will be the consequences?' The level of risk is estimated by the severity of consequences, frequency of occurrence and the correctability (correctability is also referred to as detectability). Correctability represents the probability that the failure during a process can be detected and subsequently can be corrected. The correctability is expressed in the following formula:

$$\text{Risk} = \text{severity} \times \frac{\text{Probability of failure}}{\text{Correctability of consequences}}$$

With this formula a Risk Priority Number (RPN) is calculated; here a risk matrix, discussed in paragraph 2.5, is often used. A RPN above a certain value indicates the need for actions or control measures. Hereby a distinction is made between preventive and corrective measures. Preventive measures aim at reducing the probability of failure. Corrective measures aim at improving the level of detection of failure and mitigation of the consequences. The choice for either preventive or corrective measures, depends on the practical possibilities, the costs of measures and the availability of resources. Once a preventive or corrective measure is implemented, a new RPN is calculated which makes it clear whether the risk is sufficiently reduced. In order to control a risk, it should be then ensured that the corrective and/or preventive measures are correctly and consistently applied. Corrective measures are useful only when the occurrence of failure is timely detected, preferably as a part of the ongoing process. Detection of failure can be comparatively simple – as the power failure at railway crossing results in automatic closure of the barricades; the probability of detection of power supply failure is thus very high – in other cases specific, statistically valid monitoring programmes are necessary in order to be able to indicate certain failure - for example, in case of radioactive fallout or a microbiological investigation of aseptic filling processes.

FMEA is a good method to investigate and eliminate the key causes of failure. If there are complex failure mechanisms in many underlying subsystems, then, use of a fault tree analysis is more appropriate.

2.4 The Bow Tie principle

The Bow Tie principle is a widely accepted method to analyse and assess risks. The popularity of this method is especially due to the use of diagrams that visualize the relationship between causes, consequences and control measures. The name of this method is derived from the shape of the diagram which resembles a bow tie. A Bow Tie diagram illustrates how hazards are controlled and how the associated risks are reduced. This chapter describes the general principles, history and practical application of this method.

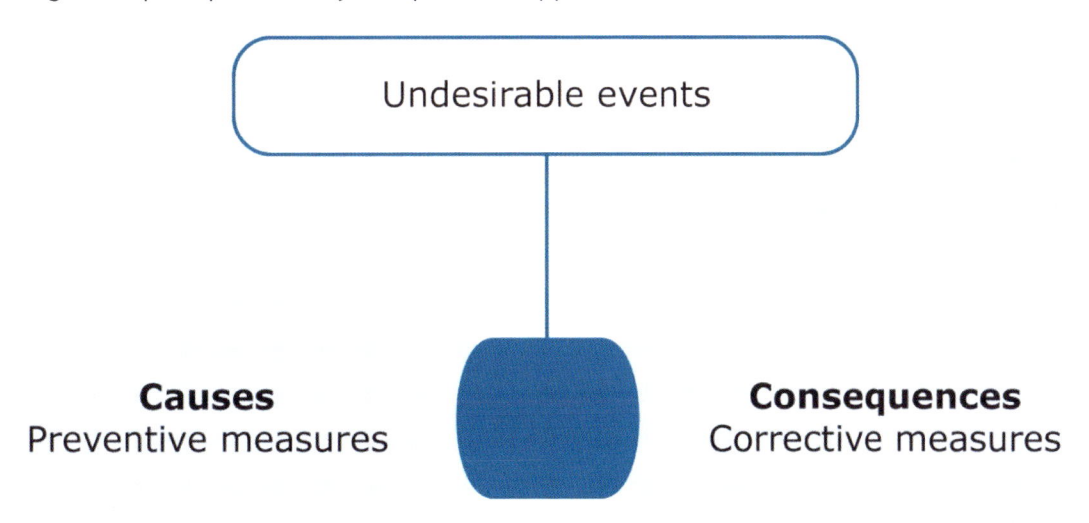

Figure 2.1 Bow Tie diagram

The Bow Tie principle is an amalgamation of two methods: fault tree analysis and the event tree analysis. The centre of the diagram represents a undesirable event, see figure 2.1 (Bow Tie diagram). In terms of the Bow Tie approach this is called the 'top event'. The left side of the diagram displays a fault tree analysis in which different causes, or threats of an undesirable event are examined. The right side of the diagram represents an event tree analysis in which different consequences of an undesirable event are examined. The fault tree displays preventive measures that deal with the causes of the undesirable event. The event tree on the right side displays corrective measures that mitigate the consequences whenever the undesirable event occurs. The strong point of this diagram is that it uses little text and is comprehensible even for the non-specialists.

The exact origin of Bow Tie approach is unclear. It is believed to be initiated from the cause and consequence diagrams used in the 1970ties. The method was mentioned for the first time in 1979 by Imperial Chemical Industries where it was used to analyse the safety in gas and oil industry. In early nineties, Royal Shell-group further developed and integrated this method in its companies. In the 90s, the method became a standard in many branches of the industry. Meantime, the use of this method has spread in many different fields such as cost control, environmental protection, personal safety, safety of buildings and constructions, in the medical world and in the political area.

The first step in drawing a Bow Tie diagram is to choose an 'undesirable event' or top event, and this is the most difficult step. The difficulty is that, in case of accidents there is usually a series of successive events. A food infection, for example, can be caused by a restaurant employee infected by *Salmonella* bacteria. Due to poor personal hygiene, the employee consequently transfers these bacteria to food which is kept too long at a high temperature and eventually consumed by an old and relatively weak consumer. All these different events in this series generally can be designated as 'undesirable events'. Depending upon the choice of the top event, a situation in one case is considered as cause and in other case as consequence of that top event. This also means that the choice of the top event determines whether an undertaken measure should be considered as a preventive measure or as a corrective measure. For example, a recall by a food producer who executes the recall, is considered as a corrective measure which aims at further reduction of the consequences caused by an error in the production (the production error is the top event for the producer). The consumer will interpret the same recall as a preventive measure that prevents a health injury (health injury is the top event for the consumer). Therefore in a risk analysis, the measures can be indicated as corrective as well as preventive, depending upon the situation to which the measure is related. The choice of an undesirable event is ultimately arbitrary and often a process of trial and error. Just as in FMEA, a risk matrix is also used in the Bow Tie approach. The Bow Tie method identifies the **possibility** of causes and consequences, while the risk matrix is a tool to assess the **probability** and severity of causes and consequences.

An interesting element in Bow Tie method is the so called 'escalation factor'. Escalation factors are situations or events which could lead to the failure of the undertaken measures. In for example a recall, an escalation factor could be that in the advertisement to warn the consumers, a wrong traceability code is communicated. Another example: hygiene rules are implemented as measures to prevent contamination but these rules are not communicated to a new employee. Escalation factors can occur in both preventive as well as in corrective measures. Escalation factors are in turn controlled by the measures which are referred to as 'escalation control'. An example of a measure to control an 'escalation factor' is the calibration of measuring instruments. More examples are cited in table 2.1.

Table 2.1 Examples of measures, escalation factors and escalation control.

	Escalation factor	**Escalation control**
Preventive measure: Maintenance	Backlog in maintenance	Internal audit
Preventive measure: Daily cleaning	Low concentration of cleaning agent	Conductivity measurement of cleaning liquid
Corrective measure: Recall	Incomplete traceability data	Periodic traceability test
Corrective measure: Procedure for emergencies	Key officials unavailable	Supplementary key official

The concept of escalation factors points out that failure(s) can occur in the measures undertaken to reduce risks. In other words, failure can occur in preventive and corrective measures. The risks of failure in preventive or corrective measures can be reduced by a subsequent stratum of preventive and/or corrective measures. Thus this creates a system of safety measures upon safety measures upon safety measures, see figure 2.2. The art of risk analysis is the ability to look through these different layers of safety and to identify at which levels different measures are active. The Bow Tie diagram is thus a user-friendly tool thanks to its visual presentation. An example of safety measure upon safety measure – or – perhaps failure upon failure - is cited in the example – *Petten two narrow escapes*.

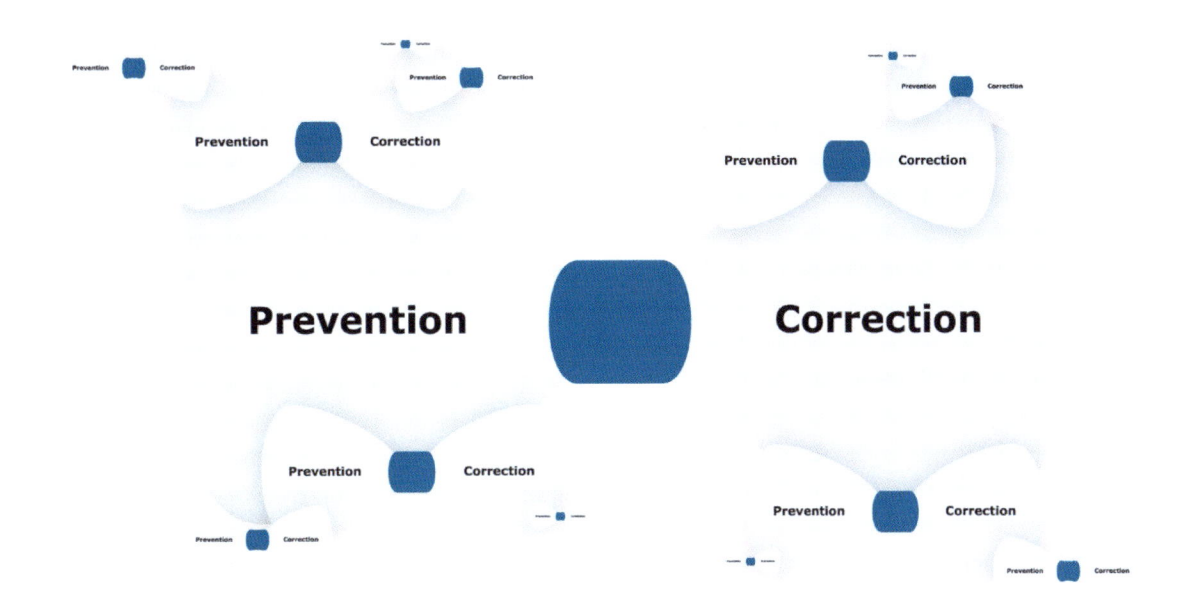

Figure 2.2 Bow Tie principle: measures upon measures

Example
"Petten: two narrow escapes"
The nuclear reactor in Petten (The Netherlands) has not once but twice escaped a meltdown. So says Frans Saris, a former director of Netherlands Energy Research Foundation (ECN), revealing a near nuclear disaster in his book Darwin Meets Einstein. Saris describes in his book how, in November 2001 a nuclear disaster had threatened when during a power failure, the primary and the emergency cooling system proved to be defective. Also the last resort, convection cooling, failed as the operator failed to open the valves due a bad lighting situation. A catastrophe was avoided as the energy supplier restored the power supply within few minutes and thus the primary cooling system was functional again.
Saris adds to these revelations that soon after, there was yet another almost-a-disaster. Against regulations, the reactor was started immediately after the first incident. 'First of all the contingency repairing should have been done. That was not the case. Because the supplier also had done insufficient repairing, the power supply failed again. Also in this case it was a narrow escape', said Saris.
(Source: Koen van Eijk, Parool-Binnenland, Saturday November 28, 2009.)

For the analysis of existing safety management systems, the Bow Tie principle is a more appropriate method than FMEA. FMEA particularly looks at the necessity to implement additional measures on top of the existing situation. The Bow Tie principle requires that the primary problem and primary causes are defined and thus insists for analysis of the already existing safety system. This is also true for HACCP based analysis that should identify, out of a mixture of already existing measures, which measures are crucial for safety of a product.

2.5 Risk matrices
Risk matrices are used in Failure Mode and Effect Analysis and in the Bow Tie approach. Risk matrices are available in different forms, from very quantitative models where severity and probability are expressed in absolute numbers, to quantitative forms where severity is expressed in terms of less severe to catastrophic and probability varies from nil to very likely.

Severity										
10	10	20	30	40	50	60	70	80	90	100
9	9	18	27	36	45	54	63	72	81	90
8	8	16	24	32	40	48	54	64	72	80
7	7	14	21	28	35	42	49	54	63	70
6	6	12	18	24	30	36	42	48	54	60
5	5	10	15	20	25	30	35	40	45	50
4	4	8	12	16	20	24	28	32	36	40
3	3	6	9	12	15	18	21	24	27	30
2	2	4	6	8	10	12	14	16	18	20
1	1	2	3	4	5	6	7	8	9	10
	1	2	3	4	5	6	7	8	9	10

Probability

Figure 2.3 Quantitative risk matrix

Figure 2.3 shows a quantitative risk matrix which is classified in three risk categories: high, medium and low. Numbers above or equal to 45 are reflected as high risks. Numbers above or equal to 20 and less than 45 are considered as medium risks. Numbers less than 20 are marked as low risks. A quantitative approach in this matrix reveals a circular shell pattern where the core of high risks is surrounded by shells of low risks. The meaning of different risk levels could be as follows:

- High risks: the risk is unacceptable. The activity must be stopped and may only be started when the risk is reduced to a medium or low risk level.

- Medium risks: the risk is acceptable on certain conditions. This means that the activity may continue as long as the conditions are fulfilled. The conditions should include that the risk is communicated to users and that there is an ongoing effort to gain further reduction of the risk.

- Low risks: the risk is deemed to be acceptable. The activity can be continued.

The risk levels high, medium and low must be chosen in such a way that low risks do not adjoin to high risks. The rationale for this is, that when a low risk adjoins a high risk, some small change in value for probability or severity could lead to a shift from an intolerable situation to a situation in which needs nothing to be done. This should be avoided.

Despite the popularity and broad distribution of risk matrices, little research has been done on its usability. This research indicates that the use of risk matrices do not always lead to the correct decisions. Risk matrices therefore should be considered as a supportive tool and must be used with caution. A lot of scientific work needs to be done before there will be anything like a *validated* risk matrix. It should be noted that the limitations of risk matrices are not so much linked to the matrices themself, but to the limitations of the formula: Risk = probability × severity. A risk matrix is simply a graphical representation of this formula. The formula is attractive due to its simplicity, however, it appears to be unable to penetrate the complexity of causes and consequences. So far, there is no scientific research concerning the role of risk matrices related to food safety. The only official publication contemplates the use of a risk matrix as part of a HACCP training on the website of FAO (Food Quality and Safety Systems – A Training Manual on Food Hygiene and the HACCP System). See also figure 2.4. The respective matrix structure corresponds with the shell-pattern displayed in figure 2.3.

	High	Satisfactory (negligible)	Minor	Major	Critical
Likelihood of occurence	Medium	Satisfactory (negligible)	Minor	Major	Major
	Low	Satisfactory (negligible)	Minor	Minor	Minor
	Negligible	Satisfactory (negligible)	Satisfactory (negligible)	Satisfactory (negligible)	Satisfactory (negligible)
	Very low	Very low	Low	Medium	High

Severity of consequences

Figure 2.4 Risk matrix of FAO

Chapter 3: Preparation of hazard and risk analysis

To prepare a hazard and risk analysis, a HACCP team is assembled. The first task of the team is to gather the required information. This chapter provides an overview of the sequential steps in the application of hazard and risk analysis and subsequently discusses the collection of information regarding products, raw-materials and processes.

3.1 Introduction

The principles of the HACCP system are incorporated as an annex in the *General Principles of Food Hygiene* since 1993. The aim of HACCP is to identify hazards and to implement measures to control them in order to ensure food safety. HACCP can be applied to the entire chain from primary production to final consumption. The implementation, wherever possible, should be supported by the scientific evidence regarding risks to the health of consumers.

The complete title of the annex of the *General Principles of Food Hygiene* is '*Hazard Analysis and Critical Control Point (HACCP) System and guidelines for its application*'. As the title suggests the annex is divided in two parts. The first part consists of an introduction, the definitions and the principles; the second part is a guideline for the application of HACCP. In the sequel of this book the second part will be referred to as 'the HACCP guidelines'. These HACCP guidelines contain the sequence for application of HACCP as shown in table 3.1.

Table 3.1 Sequence for application of HACCP system

Step 1		Assemble a HACCP team.
Step 2		Describe the product.
Step 3		Describe the intended use.
Step 4		Construct flow diagram
Step 5		On-site confirmation of flow diagram.
Step 6	Principle 1	Conduct a hazard analysis.
Step 7	Principle 2	Determine the Critical Control Points (CCPs).
Step 8	Principle 3	Establish critical limits for each CCP.
Step 9	Principle 4	Establish a system to monitor control for every CCP.
Step 10	Principle 5	Establish corrective measures which must be undertaken whenever the monitoring indicates that a CCP is not under control.
Step 11	Principle 6	Establish verification procedures to confirm the effective implementation of food safety system.
Step 12	Principle 7	Establish documentation regarding all procedures and records which are applicable to these principles and their use.

Chapters 3, 4 and 5 of this book are based on the sequence for application. Steps 1 to 5 are considered as preparation of the hazard and risk analysis and are discussed in chapter 3. The principles included in steps 6 to 12 are discussed in chapter 4. The practical application is discussed in chapters 5 and 6.

Section 3.2 of this chapter deals with the assembly of a HACCP team. One of the tasks of a HACCP team is to verify whether the prerequisite programs are implemented; this is dealt with in section 3.3. Steps 2 and 3 concerning the product and the intended use are considered together in section 3.4. Section 3.5 deals with steps 4 and 5 regarding flow diagrams and their confirmation.

3.2 The HACCP team

A successful HACCP team demands support, commitment and a thorough involvement of management and employees. For this it is necessary that all employees are aware of the consequences of unsafe food and have knowledge regarding the causes and possibilities to eliminate or control those causes. Food safety training is therefore essential for all the employees including managers, supervisors, buyers, sellers, product manufacturers and operators.

Unsafe food is a comprehensive and complex problem and the development, implementation, maintenance and optimization of a food safety management system demands continues attention. It therefore deserves a recommendation for formation of a permanent HACCP team which is well trained in the principles and contexts of HACCP. This team will comprise of a number of key members such as quality manager, manager and the heads of production departments, supplemented with non-permanent members from different production departments and staff. A legal obligation to train the HACCP team is incorporated in chapter 12 of Annex II of European Regulation 852/2004. It also includes an obligation to train production employees. In practice there can be three distinct training levels:
1. a basis food safety course for all employees;
2. an advanced food safety course for the employees who have a managerial role or who are members of the HACCP team or who are responsible for controlling CCPs;
3. a comprehensive and profound HACCP-training for the key members the HACCP team.

It is recommended to develop an ongoing training programme for employees in a company. The HACCP guidelines state that training programmes must be regularly reviewed and updated. There must be a systematic approach to ensure that people involved in food processing remain aware of all the procedures implemented for safety and quality of the products. Development of an effective HACCP plan, if necessary, demands a multidisciplinary approach including expertise in the fields such as veterinary science, microbiology, health sciences, food technology, chemistry and mechanical engineering. A food company must ensure the availability of sufficient specific knowledge and expertise. Wherever this expertise is not available within the company then advice from external experts shall be implicated. External expertise can be from the sources such as trade and product organisations, independent experts, public authorities, scientific literature and hygiene codes.

As described in step 1 of the sequence for application of HACCP, one of the primary tasks of a HACCP team is to determine the scope. The scope describes which part of the food chain is the subject of the food safety management system and indicates on which categories of hazards the system will be focused. The later implies that a HACCP plan could limit its focus on a specific hazard or category hazards, for example, pathogenic bacteria. European Legislation, however, demands that all relevant hazards should be incorporated. Determination of the scope requires clear agreements regarding responsibilities of a company in relation to suppliers and clients. For example, it must be clear who is responsible for the choice of the company that provides transportation of raw materials or end products. The person responsible for this choice must include the assessment of transport in the hazard and risk analysis and measures to control the transport should be incorporated in the procedures of the food safety management system.

The next task for the HACCP team comprises of verification and if necessary implementation of measures in the prerequisite programme. This is discussed in detail in section 3.3.

3.3 Prerequisite programmes
The introduction of the HACCP guidelines states that prior to the application of HACCP, it should be ensured that the prerequisite programmes (as described in chapters 1 to 10 of the *General Principles of Food Hygiene*) are applied. These programmes must be well introduced, fully operational and verified in order to facilitate successful application and implementation of HACCP. This assertion leads to the notion that the prerequisite programmes and HACCP are two separate approaches where measures belonging to the prerequisite programmes cannot be an element of HACCP and vice versa. Closer contemplation, however, shows that the prerequisite programmes and HACCP are highly intertwined: section 5.1 of the prerequisite programme shows that the application of HACCP is in fact a prerequisite.

In overview one could say that the prerequisite programme is a complete package of measures that are necessary to control the safety of food. The majority of measures are aimed at preventing contamination with additional measures such as temperature monitoring, recall, traceability and training of employees. Also see section 1.5 of this book. By implementing the prerequisite programme, a company primarily controls food safety, if it was not for specific hazards or control measures which are not covered by the prerequisite programme. HACCP aims at understanding the contribution that the various measures in the prerequisite programme offer to control the risks. Based on this, HACCP provides for additional measures to reinforce the risk control and especially to enable food processors to demonstrate they control the safety of their products.

As mentioned above, prerequisite programmes and HACCP are highly intertwined and cannot be considered as two totally separated approaches. In practice, prior to the implementation of HACCP, it is sensible for food companies to ensure that the relevant requirements of the prerequisite programme are implemented. Where the prerequisite programme offers room for interpretation (through the use of expressions such as 'if necessary' and 'wherever applicable'), the HACCP team should discuss the hazards and risks related to the specific requirement in the prerequisite programmes. However, the HACCP is complex and extensive while the application of prerequisite programme is relatively simple and compact. The prerequisite programme in any case ensures a basic foundation for the control of food safety. In chapter 4, the relationship between prerequisite programmes and HACCP will be discussed in detail and also include the operational prerequisites in ISO 22000.

3.4 Description of products

Knowledge of the composition and properties of raw materials and products is of great importance for the application of HACCP. Hence, the specifications of raw materials must be available and a specifications of products should be made. This includes all relevant information regarding, for example, composition, physical and chemical properties such as Aw-value and pH, treatments for eradication or inhibition of micro-organisms (such as heat treatment, freezing, pickling, smoking, etc.), packaging, shelf-life and storage conditions, and the ways of distribution. For the companies handling a variety of raw materials and/or products, for example, during the production of meals, it is advisable to create a number of group of raw materials and/or products with similar properties, treatments or methods of preparation before developing a food safety management system.

The description of a product must include its intended use. The intended use must be based on the expected use of the product by the final user and/or consumer. Wherever applicable vulnerable target groups, like babies or sick people, should be incorporated in the intended use. Moreover, potential misuse of products should be taken into consideration. The following example illustrates the importance of incorporating potential misuse.

Example
"Potential misuse: raw cookie dough "
A producer of ice cream, uses raw cookie dough in one of its products and came across the presence of Salmonella bacteria in the dough. Hereupon the flour supplier was held liable for the presence of Salmonella in wheat flour. However, for the flour supplier it is obvious that the absence of Salmonella or any other pathogenic bacteria by no means can be guaranteed. Wheat, like all other grains, are cultivated in open air and come in contact with contaminants from air and soil and also via animals such as birds and mice. Presence of Salmonella bacteria is usually not a problem as flour is used in products that are baked or cooked, as a result of which the bacteria are killed. Based on a rapid risk analysis the supplier decided to include the following notification in the specification of the wheat flour: 'wheat flour, only for use in products to be baked or cooked'.
End of example

There are no legal requirements or commonly accepted standards for drafting specifications. Besides aforementioned characteristics, it is important for hazard and risk analysis that specifications focus on specific legal requirements. Wherever applicable specifications must include information regarding hazards that are subject legal requirements.

3.5 Flow diagrams

Flow diagrams should be developed to create an overview and get an insight in the process in which a food product is produced. By a flow diagram, it can be determined at which stages of the process contamination can occur or a hazard can develop, and at which stages control measures are taken or should be taken.

There are no concrete criteria that indicate the level of detail to which a flow diagram should be drawn. It is recommended to keep the diagram as simple as possible as to retain an overview of the process. It can be helpful to highlight control measures and critical points in the flow diagram. Further details can be optionally included in an explanatory process description. When during hazard and risk analysis the need for more details arises, further details in the flow diagram or process description may be included.

In order to ensure that the flow diagrams and process descriptions comply with the real practice, they must be confirmed by the operators who are familiar with the course of processes in different stages and circumstances of production.

Chapter 4: The principles of hazard and risk analysis.

This chapter is mainly concerned with the causes of unsafe food and the different measures to deal with these causes. This chapter also discusses the principles and definitions of HACCP and the concept of the operational prerequisite programmes (OPRPs) in ISO 22000. These concepts are blended with the principles of the Bow Tie approach.

Definitions of HACCP

Hazard: A biological, chemical or physical agent in, or condition of, food with the potential to cause an adverse health effect.

Control measure: Any action or activity that can be used to prevent or eliminate a food safety hazard or reduce it to an acceptable level.

Critical control point (CCP): A step at which control can be applied and is essential to prevent or eliminate a food safety hazard or reduce it to an acceptable level.

Step: A point, procedure, operation or phase in the food chain, including raw materials, from primary production to ultimate consumption.

Critical limit: A criterion which makes a distinction between acceptable and unacceptable.

Deviation: Failure to meet critical limits.

Monitoring: The act of conducting a planned sequence of observations or measurements of control parameters to assess whether a CCP is under control.

Corrective action: Any action to be taken when the results of monitoring at the CCP indicate a loss of control.

Verification: The application of methods, procedures, tests or other evaluations, in addition to monitoring, to determine compliance with the HACCP plan.

Validation: Obtaining evidence that the elements of the HACCP system are effective.

(Source: The General Principles of Food Hygiene)

4.1 Introduction

The principles of HACCP are included in the step by step plan given in chapter 3. Hazard analysis is the first principle of HACCP. This principle is implemented in step 6 of this plan. According to diagram 1 from HACCP guideline this step comprises of the following components:
1. make a list of all potential hazards;

2. perform a hazard analysis;

3. contemplate control measures.

In the first component, the HACCP team prepares a list of all hazards which could reasonably appear at each step described in the scope. The team then performs a hazard analysis in order to identify those hazards that must be controlled. In conducting the hazard analysis, wherever possible the following should be included:
- the likely occurrence of hazards and severity of their adverse health effects;

- the qualitative and/or quantitative evaluation of the presence of hazards;

- survival or multiplication of micro-organisms of concern;
- production or persistence in foods of toxins, chemicals or physical agents; and,
- conditions leading to the above.

Subsequently, it must be considered which control measures can be applied to control different hazards. It is possible that several control measures are necessary for controlling one specific hazard, whereas a number of hazards can also be controlled by one particular control measure.

The above methodology of HACCP has resulted in hazard analysis in which a HACCP team indicates the possible occurrence of microbiological, chemical or physical hazards for each and every process step. This process is repeated at every subsequent process step. This 'process step oriented' approach has some disadvantages. HACCP, in fact, is a *hazard analysis* and not a *process step analysis*. The name has already suggested that the analysis should focus on hazards and not on process steps. It is not just about a the ability of a companies to demonstrate that certain processes or process steps are controlled; a food safety management system should enable a company to prove that a particular hazard is controlled. This is especially important when a company is confronted with incidents in which consumers suffer serious adverse health effects. In such a situation, the attention is focussed on that part of the food safety management system that is related with the hazard of concern. For example, when the incident concerns pieces of glass in a product then one is interested in the hazard and risk analysis of physical hazards. It is undesirable when, in that same analysis 70% of the content is dedicated to other hazards. In the worst scenario the hazard analysis offers an insufficient overview that undermines trust in the food safety management system of the company. A second disadvantage of the 'process step oriented' approach relates to the functioning of the HACCP team. In practice, especially for microbiological hazards, a need for control measures at a certain process step can only be assessed in respect of the entire process. Therefore, for a HACCP team it is easier and more convenient when first the analysis of the entire process is done, for example, on microbiological hazards, then (in random order) on chemical hazards, physical hazards and on allergens. This of way working prevents the repetition of the discussion about a certain category of hazards at each and every process step.

This book, therefore, presents an approach that creates a distinct hazard and risk analysis for each category of hazards. This highly benefits the quality and the legibility of the hazard and risk analysis. Nevertheless, there is also a deliberation for aggregating the different hazards per process step. This can be, for example, serve the production people who would like to have an overview of different hazards, which can occur at a certain step. For people involved in the purchase of raw materials, it is good to know about the different hazards related to a specific raw material. Ideally, a hazard and risk analysis could be sorted with software. In one case control of a certain hazard can be made perceptible by selecting apt process steps and raw materials. In another case, it is possible to make different hazards per process step or per raw material perceptible.

For a proper understanding of hazard and risk analysis, this book diverts it's attention from process steps to hazards. A thorough understanding of food safety starts with knowledge of adverse health effects caused by unsafe food and hazards involved. A hazard and risk analysis, therefore, should start with the creation of a list of relevant hazards that should be controlled by the food safety management system.

4.2 List of relevant hazard

To create the list of relevant hazards, the HACCP team should investigate which hazards (hazardous agents or substances) were found by consumers in their food and in which products these hazards were found. The HACCP team should then decide which of these hazards would potentially reach the plate of consumers via the particular products of their company. These are the relevant hazards. The result is a list indicating whether or not a hazard is relevant to the products of the company. The severity and probability of the relevant hazards are examined. In the hazard and risk analysis, executed in a following step, it is decided where in the process control measures for these relevant hazards must be implemented. A difference in the 'list of relevant hazards' and the 'list of potential hazards' as described in the HACCP guidelines is that the latter is linked to the various steps in the process while the former focuses at the product and the consumer and is independent of the process. A significant advantage of the former is that the consumers and their health are the focus of the food safety management system and not the process of the producer.

Information to create a list of relevant hazards is available from various sources. The principal source is the legislation wherein hazards are mentioned often in combination with products or raw materials. Next to this, companies can acquire their own complaint history and that of their peers. Furthermore, a lot of information about hazards is available in literature and on the internet. Remarkable sources on the internet are the sites of RASFF (Rapid Alert System for Food and Feed), EFSA and the site of New Zealand Food Safety Authority. Interesting scientific books and reports are written by ICMSF (International Commission for Microbiological Specifications for Foods).

A remark regarding the list of relevant hazards is that next to relevant hazards it also includes some hazards that are not relevant. In order to prevent the endless lengthening of the list containing irrelevant hazards, it is advised to include only those irrelevant hazards which appear quite commonly in the food sector. Hazards such as *Listeria monocytogenes* and *Salmonella* should be included in every list of relevant hazards due to their relatively common occurrence and severity of health effects.

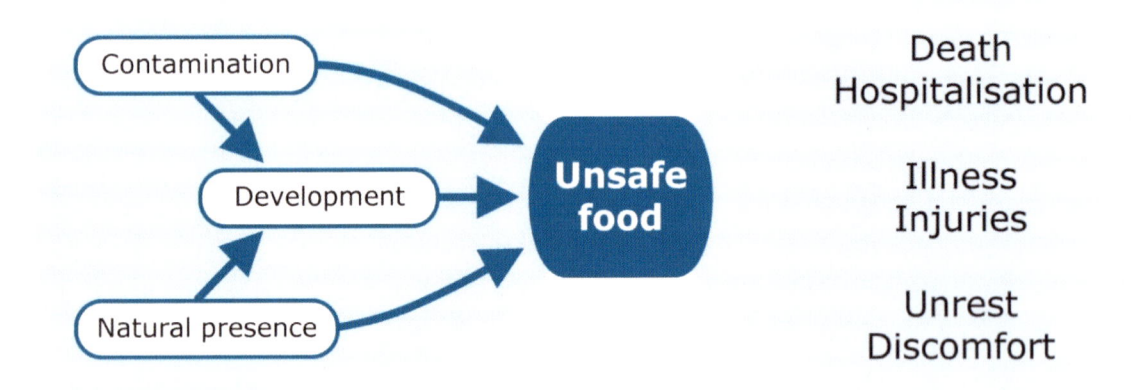

Figure 4.1 Basic Food Safety Bow Tie diagram

Figure 4.1 (Basic Food Safety Bow Tie diagram) present a concept to support the creation of the list of hazard. The figure presents the primary causes of unsafe food. In this basic Bow Tie diagram the undesirable, or top event, is 'offering unsafe food to consumer'. The choice for this top event is arbitrary. A process of trial and error pointed out that a Bow Tie diagram constructed on this top event gives the best overview of the problem of unsafe food.

The right side of the Food Safety Bow Tie shows the possible consequences which are reflected as categories of adverse health effects. On the left side the potential threats are indicated. In case of unsafe food the threats consist of the various ways through which hazardous agents can end up in food. In HACCP these hazardous agents substances are referred as 'hazards'. The definition specified in the *General Principles of Food Hygiene* is as follows:

Definition of hazard
Hazard: A biological, chemical or physical agent in, or condition of, food with the potential to cause an adverse health effect.

It should be noted that, regarding this definition from 2003, the category 'allergens' is not specified. However, allergens can also be interpreted as chemical hazards. The definition indicates that next to agents, a second type of hazard exists called 'conditions'. In the text below hazards are discussed first as 'agents'. Hazards that could be referred to as 'conditions' will be addressed later on. Hazardous agents can end up in food by three different ways. These ways have been indicated as threats on the left side of the Basic Food Safety Bow Tie diagram.

First threat: Hazardous agents as naturally presence
Certain hazardous agents are inextricably associated with 'natural character' of a primary raw material. Examples are bones in fish, seeds in fruits, toxins in plants (phytotoxins) and allergens in peanuts and wheat.

Second threat: Contamination
Contamination refers to hazardous agents that end up in raw material or food from the environment, including process equipment and packaging materials, in which food is produced. The different sectors of the food chain recognize their typical forms of contamination, for example, environmental contaminants are typical for the primary sector and lubricants are typical contaminants for the industry. Viruses perhaps play an important role during preparation, while contamination due to moulds (Fungi) and bacteria occurs throughout the chain.

Third threat: Development of hazardous agents
Besides 'natural facts' and 'contamination' there is a third type of threat: development of hazardous agents. There are two types of development:
1 - development of hazards that are present due to contamination.
An example is the production of toxins as a consequence of fungal or bacterial contamination. A second example is when bacteria can grow from a low and acceptable level up to numbers, which can be certainly harmful to health.
2 - development of hazards as reaction products of agents that are naturally present.
The formation of acrylamide is an example of this type of development. Acrylamide is formed as a product from a reaction between reducing sugars and amino acids. This reaction occurs during frying of potatoes. Another example is the formation of 3-monochloro-1,2-propanediol (3-MCPD) during acid hydrolysis of vegetable proteins for production of flavouring agents and soy sauce.

In addition to hazard as agents, the definition of hazards also refers to 'a condition'. The *General Principles of Food Hygiene* are unclear about what should be considered to be such a 'hazardous condition'. The formal HACCP system doesn't offer any explanation or example. A first possible interpretation is that a 'condition' is comprehended as a condition that is capable of causing negative health effects. Examples could be: coffee which is served too hot, croquettes that explode due to internal steam formation, deep-fry products causing splashing of oil, exploding bottles of soft drinks. This type of condition is often very product specific and shall not be discussed here further. However, it is recommended to incorporate them appropriately in the food safety management system. See the example 'Dangerous olive oil aerosol dispenser off the shelves'.

Example
"Dangerous olive oil aerosol dispenser off the shelves"
The olive oil aerosol dispenser of Brand X is dangerous. This has been decided by the Food Safety Authorities in response to our complaint. The importer has withdrawn the aerosol dispenser from the market. Our complaint is based on a rapid test which shows that the nozzle of an aerosol dispenser is difficult to point downwards. Therefore, the chances are that the olive oil spray ends up in an adjoining blazing gas burner, with a blowtorch flame as a consequence. We hope that the food industry has now permanently abandoned the evil idea of providing aerosol dispenser with inflammable content which one would use nearby a blazing gas stove or other heat source. Around two years ago we had already taken steps against another aerosol dispenser with olive oil.
(Source: Consumentengids, the Netherlands, March 2003)
End of example

A second possible interpretation is to regard a 'condition' as a situation or event, which could lead to the presence of hazardous agents. For example, a leak in a package could lead to contamination or a failure of a chiller causing growth of microorganisms. These are conditions which are indirect causes of an adverse health effect. The approach of HACCP in this book exclusively focuses on hazards in the context of hazardous agents. Conditions which can lead to the presence of hazardous agents are discussed in paragraph 4.4 as failures in control measures.

4.3 Control measures

In chapter 4.2 the HACCP team created a list of the relevant hazards that at an unguarded moment could end up in the products of a company. The HACCP guidelines indicate that for these hazards, wherever possible an assessment should be done of the probability of occurrence and the severity of adverse health effects. Assessment the probability and severity, or risk, will be discussed in Chapter 5. The sequel of this chapter deals with different types of measures, which can ensure the reduction of the probability of consumers being affected by a hazard. It will look at the difference between control measures and monitoring procedures and the position of these measures in the Food Safety Bow Tie.

The definition of control measures from the *General Principles of Food Hygiene* is presented as follows:

Definition: Control measure:
Any action or activity that can be used to prevent or eliminate a food safety hazard or reduce it to an acceptable level.

Control measures thus are the actions and activities with which the presence of hazardous agents in food can be prevented or reduced. Eliminate can be regarded as a specific type of reduction. Different types of control measures are discussed in the sequel of this paragraph. An overview has been given in Figure 4.2.

4.3.1 Prevention of hazards

There are two possibilities for the prevention of hazards:
        a. Excluding or limiting contamination; and

        b. Excluding or limiting development.

Excluding or limiting contamination
Measures such as personal hygiene, pest control, cleaning and disinfection and maintenance of buildings and equipment aim at controlling hazards at the source. These measures lower the contamination pressure whereby the probability of product contamination is reduced. In some situations, for example *Salmonella* in ready to eat raw vegetables, it is important that the primary contamination sources and the corresponding vectors are thoroughly analysed in order to undertake adequate measures. A possibility to exclude or limit contamination is to protect the product with a package or by processing and in closed equipment. Food packaging primarily focuses on ease of handling and attractive presentation of the product but at the same time food packaging can be a control measure to prevent contamination. See the example 'Botulism through can of Salmon'. It is also possible to restrain the source of contamination, for example, by plastic film around fluorescent lamps and on windows.

Example
"Botulism through can of Salmon"
Four pensioners in Birmingham were affected by Type E Botulism in 1978. After eating a can of Salmon on Sunday, July 30 around 16.00 pm, the first duo started to suffer from vomiting and severe diarrhoea at around 2.00 am. After a couple of hours, they were admitted to a hospital in acute distress. There it was noticed that they suffered from blurred eyesight and had difficulties while speaking. Police found the second duo who had participated in the meal, in a desperately ailing and helpless condition. They were immediately brought to the hospital where their condition rapidly deteriorated. Around 10.00 in the morning they no longer could breathe as their muscles paralyzed. All four victims were then shifted to an intensive care unit and treated with an antidote. However, both, the first and the second duo succumbed to death after 13 and 23 days respectively.

The causes of death were breathing problems and cardiac arrest. Clostridium botulinum Type E was found in the leftovers from the can of Salmon.
It was later revealed that there was an error in the manufacture of cans, which lead to the contamination during their cooling in a plant in Alaska. It was assumed that the contamination was caused because of the production workers handling raw fish laid their wet outfits on the cooling cans for drying. This lead to post heat treatment contamination as spores of Clostridium botulinum could penetrate the erroneously manufactured cans. If it has really gone this way then this is a perfect example of consequences of cross contamination from raw to cooked food. This example also shows how important a packaging can be as a control measure for the food safety.
(Source: S.J. Forsythe & P.R. Hayes, Food Hygiene Microbiology and HACCP.)
End of example

Excluding or limiting growth
Many foodstuffs owe their microbiological safety to the fact that their composition is such that hazardous bacteria and fungi that are likely to be present will not get any opportunities to grow. Besides water content, presence of sugar, salt, acids, also other preservatives play an important role. For some products, this type of control measure is a consequence of the method of preparation like low pH for yogurt. For other products, this control measure is intentionally and actively introduced, for example, lowering the pH of mayonnaise by adding vinegar (acetic acid). The growth of microorganisms can also be slowed down or stopped by keeping food under certain conditions. Examples are gas and vacuum packaging, refrigerating and freezing.

4.3.2 Elimination or reduction of hazards
Control measures for elimination or reduction of hazards can also be divided in two different types:
   a. complete (elimination) or partial removal (reduction) of hazardous agents; and

   b. complete (elimination) or partial removal (reduction) of unsafe products.

Complete (elimination) or partial removal (reduction) of hazardous agents
There are several processes by which the hazardous agents can be removed from food or made harmless. Physical contaminants can be removed with the help of sieves and filters. Chemical contaminants can be removed by using activated coal. Bacteria and fungi can be made harmless by heat treatment. However, due to heat stability this does not apply to many toxins produced by bacteria and fungi.

Complete (elimination) or partial removal (reduction) of unsafe products
Chemical contaminants such as mycotoxins, in many cases cannot or hardly be removed from food. In a number of situations such as aflatoxin in pistachio nuts, the producers are not capable to fully prevent fungal contamination and development of toxins. The last possibility to control such a hazard is to exclude unsafe batches of raw material from the process. In fact, this is a type of screening of unsafe products or a procedure for positive release. The positive release is done on the basis of representative sample and analysis to determine which batches are safe and which are unsafe. Only those batches that meet the limits can be used as raw material for the production of food. Also see paragraph 5.3 from the *General Principles of Food Hygiene*. In fact, a metal detector or any other kind of detector is also a type of positive release of safe products. With the help of such devices all the products are analysed, and unsafe products are excluded from the process and safe products are released for further processing.

4.3.3 Food Safety Bow Tie
The control measures discussed before are incorporated in figure 4.2.

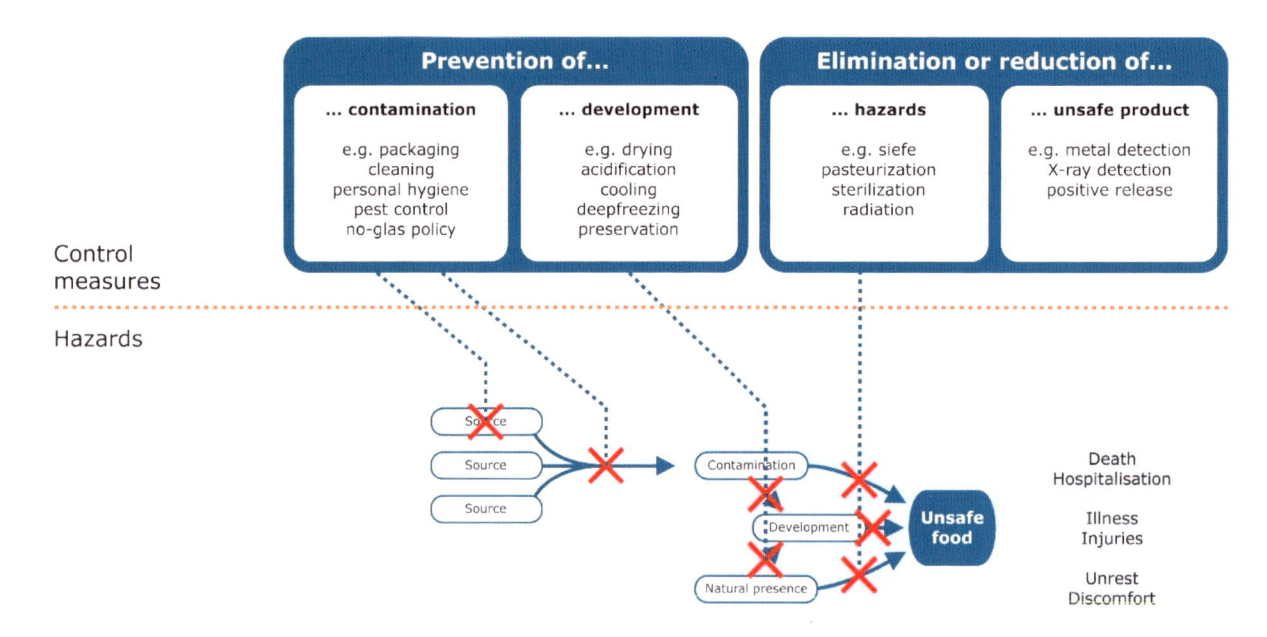

Figure 4.2: Control measures incorporated in the Food Safety Bow Tie.

Lastly, there are no control measures for preventing 'naturally existing' hazards. In some cases, these hazards can be eliminated or reduced, for example, by deboning fish or by deseeding cherries. There are no control measures for naturally existing allergens. A measure in the form of a message on the label to warn consumers with allergie is, in the strict sense of the definition not a control measure. This warning to the consumers, reduces the risk (the likelihood of allergic reactions is reduced) but the hazard is not prevented, eliminated nor reduced – the allergens will remain and the product continues to be unsafe for the allergic consumers.

Implementation of control measure is the last part of the first principle of HACCP. In the next step, the critical control points are determined in accordance with the second principle.

4.4 Critical control points

Table 4.1 presents a comparison between the definitions of control measures and critical control points.

Table 4.1 Comparison between definitions of control measures and critical points.

Control measure	Critical control point (CCP)
Any action or activity that can be used	A step (a point, procedure, operation or stage in the food chain including raw materials, from primary production to final consumption)
	at which control can be applied and is essential
to prevent or eliminate a food safety hazard or reduce it to an acceptable level.	to prevent or eliminate a food safety hazard or reduce it to an acceptable level.

A comparison between these definitions indicates that critical control points must be considered as control measures at which control can be applied and are essential. In other words, critical control points are 'controllable and essential control measures'. (Critical control points are, in fact, critical control measures and therefore, HACCP might as well be renamed to HACCM – Hazard Analysis Critical Control Measures.)

Essential control measures
The word essential in relation to a control measure means that the control measure cannot be missed for the safety of product. The term 'essential' in relation to control measures could regarded synonymous for 'critical'. Both the terms 'essential control measures' or 'critical control points' indicate that the measure is decisive for the safety of the product. Limiting contamination or growth may be essential when a hazard cannot be eliminated or reduced; the elimination or reduction of a hazard is possibly essential whenever contamination or growth cannot be prevented. Thus, for example, a high level of personal hygiene is essential during compilation of ready to eat meals, especially when these meals have no subsequent heat treatment. The elimination or reduction of pathogenic microorganisms by heating chicken is essential since contamination caused by these microorganisms cannot be excluded completely. The essentiality or criticality of a control measure relates that the probability and severity of the concerned hazard. The production process of salt has a low risk for microbiological hazards, so control measures in relation to microbiological contamination should not be designated as essential. A tool that makes a distinction between essential and non-essential control measures is a decision tree, that will be discussed in detail in chapter 5.

Once a control measure is identified as essential, the subsequent question is: is this a control measure *at which control can be applied*? The answer becomes clear after referring principles 3, 4 and 5 of HACCP. On the basis of these principles it can be put forth that CCPs:
1. are essential control measures;
2. for which critical limits must be established (principle 3);
3. which must be monitored so that any possible deviations with respect to critical limits can be indicated (principle 4);
4. so that the products which have become or remained unsafe due to these deviations should be excluded from the process (principle 5).

Based on the above discussion, it can be stated that control measures *at which control can be applied* are control measures that are feasible for monitoring and corrective actions. Considering the definitions and principles, the critical points should be regarded as 'essential control measures that are feasible for monitored and corrective actions'. The primary aim of HACCP system is thus to implement procedures to monitor and correct failure in essential control measures.

4.4.1 Deviations in control measures in the Bow Tie approach
One of the principles of the Bow Tie approach is that the measures undertaken to reduce risks, can fail. In the Bow Tie approach this is indicated as escalation. Escalation can be controlled by measures, which are designated as 'escalation control'. In HACCP a 'deviation' - or failure to meet critical limits – can be interpreted as escalation. Monitoring and corrective measures in HACCP can be considered as escalation control. Examples of failing control measures are a pasteuriser that does not reach the desired temperature, exposure of refrigerated products to the sun while being transferred from a lorry to a storage room and employees who do not abide hygiene rules.

In the context of the Bow Tie approach, failure of a control measure can be perceived as an 'undesirable event', see figure 4.3. The risk of failure of a control measure can be controlled 1) by preventive measures, which reduce the probability of failure or 2) by corrective measures that detect failure and allow corrective action. The latter is especially important for the control measures which are identified as essential, as in case of CCPs. For CCPs the risk of failure of the control measure is controlled by a monitoring system, which detects the deviation and the subsequent corrective measures will ensure that the possible unsafe products are excluded from the process.

Preventive measures
Measures to reduce the probability of deviations

Deviated control measure

Corrective measures
Monitoring procedures with critical limits and corrective measures

Figure 4.3 Prevention and correction of failing control measures

4.5 Operational prerequisites programmes

In the preceding paragraph critical control points, CCPs, are described as 'essential control measures for which monitoring and corrective action is feasible'. This paragraph deals with the 'essential control measures for which monitoring and corrective action are not feasible.

First a little bit of HACCP history. The International Commission on Microbiological Specifications of Foods (ICMSF) has played an important role in the development of HACCP. A 1988 publication of the Commission describes two types of CCPs. See figure 4.4. The differences between these CCPs are illustrated as follows: 'Some food processes include procedures where a CCP can completely eliminate one or more microbiological hazards. Such a CCP is described as CCP 1. The CCP can ensure control of hazard by frequent measurement of parameters such as temperature and time (as in pasteurization and refrigeration). It is also possible to identify CCPs which can reduce a hazard but where control cannot be guaranteed. These CCPs are described as CCP 2. Both types of CCPs are important and must be controlled. Some CCPs cannot be monitored continuously, and the control is established by periodic online or offline measurements, for example, visual inspection during evisceration of guts (intestines) in a slaughter process and checking of seam during canning of products. (Source: *Micro-organisms in Foods part 4, Application of HACCP to ensure microbiological safety and quality*, ICMSF, 1988.)

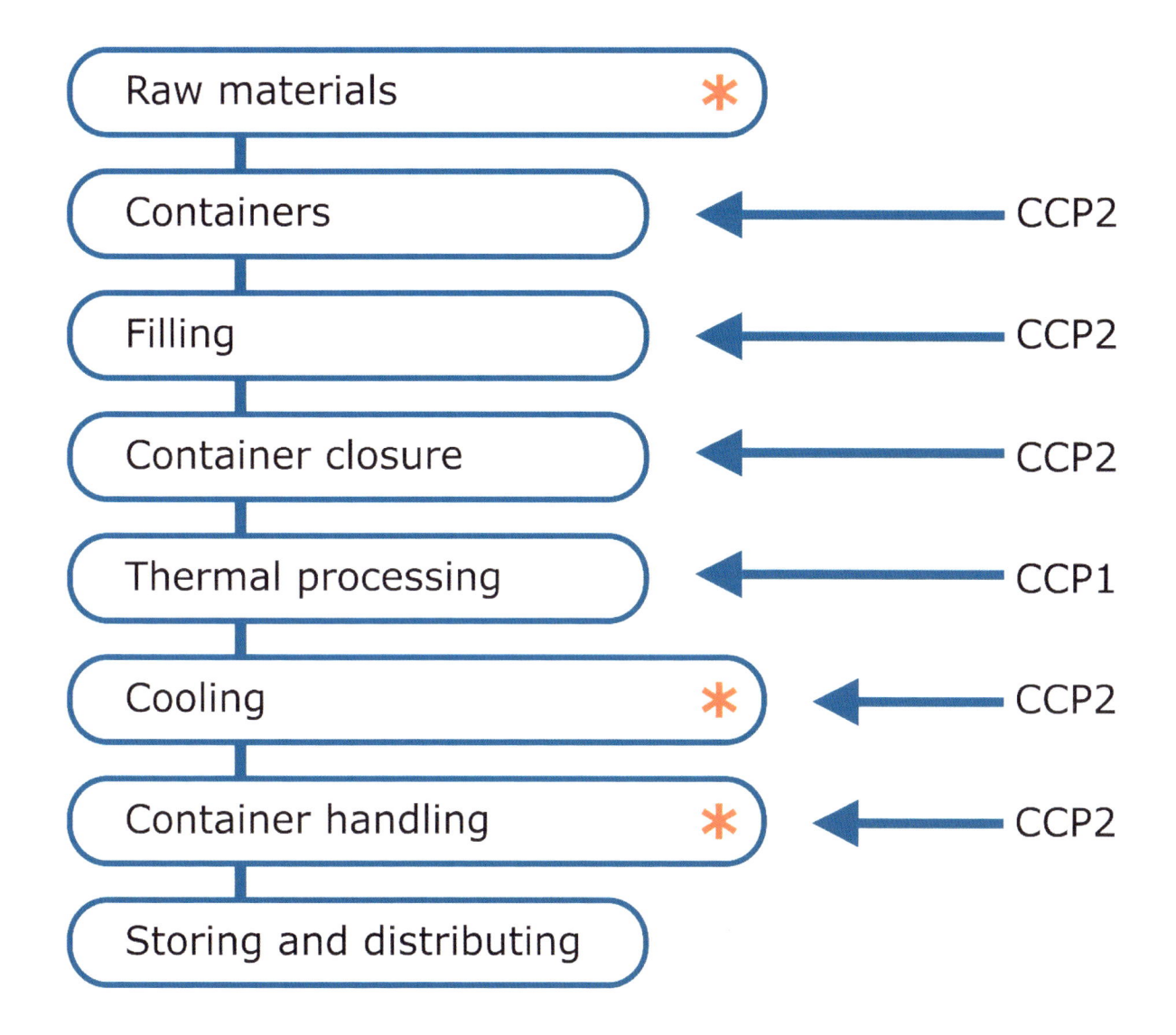

```
Raw materials                    *
      ↓
Containers                    ← ——— CCP2
      ↓
Filling                       ← ——— CCP2
      ↓
Container closure             ← ——— CCP2
      ↓
Thermal processing            ← ——— CCP1
      ↓
Cooling                    *  ← ——— CCP2
      ↓
Container handling         *  ← ——— CCP2
      ↓
Storing and distributing
```

* Indicates a site of major contamination
CCP1: Effective CCP, CCP2: Not absolute

Figure 4.4 CCP 1 and CCP 2

In the political process in which the HACCP approach was prepared to be incorporated in the Codex Alimentarius, the concept of CCP 2 met with an obstinate resistance. Certain parts of food industry refused to reveal that control could not be guaranteed at all points. This eventually led to exception of the concept of CCP 2 from the *General Principles of Food Hygiene*.

The issue of essential control measures not feasible to monitoring, however, was not eradicated. In the certification standard Dutch HACCP published in 1996, the term 'points of attention' was introduced. In the third revision of this standard of September 2002, the 'points of attention' are replaced by the 'general control measures'. The general control measure was defined here as 'a measure to control a specific part of the prerequisite programme'. The general control measure is thus distinguished from the specific control measure which is defined as a measure to control a critical control point (CCP). This approach attempts to solve the problem of the essential control measures at which control cannot be applied, by relating these control measures to the prerequisite programme and consequently, evading principles 3, 4 and 5 of HACCP. This approach is also reflected in ISO 22000. Section 7.4.4 'Selection and assessment of control measures' of ISO

22000 states that the control measures must be classified as measures which must be controlled via operational prerequisite programmes (OPRP) or as CCPs. ISO 22000 uses diverse definitions and at least seven different criteria to make a distinction between the operational prerequisite and the critical control points. These seven criteria will be further discussed in chapter 5. The definitions used by ISO 22000 are as follows:

Definitions from ISO 22000
Prerequisite programme (PRP): *Basic conditions and activities necessary to maintain a hygienic environment throughout the food supply chain which is suitable for production, handling, and provision of safe end products and safe food for human consumption.*

Operational prerequisite programme (OPRP): *PRP identified by the hazard analysis as essential in order to control the likelihood of introducing food safety hazards to and/or the contamination or proliferation (or spread) of food safety hazards in the product(s) or in the processing environment.*

Critical control point: *A step at which control can be applied and is essential to prevent or eliminate a food safety hazard or reduce it to an acceptable level.*

The definitions from ISO 22000 indicate that both OPRPs and CCPs must be considered as essential control measures. Deviations in these control measures must be prevented or wherever possible must be monitored and corrected. Also see figure 4.3. OPRPs focus on the elimination of the causes of deviation while CCPs emphasize on detection and correction of deviations. The monitoring of CCPs guarantees the control but the monitoring of OPRPs does not give such a guarantee. Table 4.2 shows that the main difference between OPRPs and CCPs is the absence of critical limits. Figure 4.5 depicts the Food Safety Bow Tie, including deviations in control measures and the subsequent relevant preventive and corrective measures.

Table 4.2 Comparison between OPRPs and CCPs

	Operational Prerequisite Programmes (PRPs – 7.5 in ISO 22000)		**HACCP – Plan (CCPs – 7.6.1 in ISO 22000)**
	The operational PRPs must be documented and each programme must include the following information:		The critical control points (CCPs) must be documented and each CCP must include the following information:
a.	food safety hazards which must be controlled by the programme	a.	food safety hazards which must be controlled at the CCP
b.	control measures	b.	control measures
		c.	critical limits
c.	monitoring procedures to demonstrate that the operational PRPs are introduced	d.	monitoring procedures to verify if critical limits are exceeded.
d.	corrections and corrective measures which should be undertaken when the monitoring indicates that no control is offered by PRPs.	e.	corrections and corrective measures which should be undertaken when the critical limits are exceeded.
e.	responsibility and competencies	f.	responsibilities and competencies
f.	registration(s) of the monitoring	g.	Registration(s) of the monitoring

To conclude this section, two examples to illustrate essential control measures at which monitoring and corrective actions are not feasible.

Example
"Japanese enjoy puffer fish"
People gladly pay a few hundred euro for a puffer fish meal as it is very delicious. Apart from its delicacy, consumption of puffer fish is also dangerous. Every year around ten Japanese drop their chopsticks after eating puffer fish. They suffer from breathlessness, convulsions, paralysis and eventually death.

There is an enough toxin in one puffer fish to kill 30 adult humans. This toxin tetrodotoxin is 275 times more poisonous than cyanide. The toxin is mainly present in the lever, gallbladder, gonads and skin. Thus careless cleaning of the fish is fatal. The intoxication symptoms are horrifying. All muscles in the body are slowly paralyzed. Finally, the lungs and heart stop. Death results after 24 hours. During this period, the victim is fully conscious. There is no known antidote for this toxin.
Several times in the course of history, consumption of this fish has been banned, but it was in vain. The fish continues to be popular. In 1958, a strict law was passed for Japan's hospitality industry. Puffer fish dish in Japanese restaurants can only be prepared by specially trained and licensed chefs. They ensure that not even a droplet of poison comes into contact with fish-meat. Despite a few exceptions, this is successful.
Chefs must follow training for three years and also after that they are often examined strictly for their capabilities. Only about 30% of the trainees finally obtain the diploma. The requirements are so stringent that the smallest hitch in the preparation keeps a candidate away from his diploma. Therefore, since 1958 there have hardly any accidents occurred in the official restaurants. The Puffer fish accidents occur at home by hobby cooks.

Explanation:
Tetrodotoxine is the hazard. Source are the viscera. The control measure is preventing contamination through careful evisceration. Any damage to the viscera indicates failure of the control measure. The occurrence of this damage is not or hardly detectable. Monitoring such as inspection during evisceration gives no guarantee of complete and proper implementation of the control measure. Training the chefs is a preventive measure which ensures reduction of probability of failure of the control measure. Unfortunately, there is no guarantee.
End of example

Example
"Pathogens in raw products"
Hazards such as Listeria monocytogenes in soft cheese made from raw milk and Escherichia coli O157H in filet American or consumption of raw steak tartare are mainly controlled by OPRPs. For such products contamination and development of these hazards must be restricted.
In the production of soft raw milk cheese hygiene during milking process is very important. This also applies to storage and further processing of milk. The risks of failure in hygiene (such as personal hygiene, cleaning and disinfection, prevention of cross-contamination) are significant and hence hygienic processing should be considered as essential. However, a practical problem is that it is impossible to define objective and measurable requirements for hygienic processing. It is certainly possible to standardise or measure some aspects (such as temperature and concentration of detergents) but a univocal, objective and consistent measurement capable of excluding all risks is not possible. The control measure hygienic processing thus should be regarded as OPRPs and not as CCPs. Cooling of milk and soft raw milk cheese can be regarded as a CCP: as critical limits can be defined for cooling (maximum 7°C) and a monitoring system can ensure and demonstrate control of the cooling. However, cooling does not guarantee control of pathogenic microorganisms as it cannot undo unacceptable contamination.
End of example

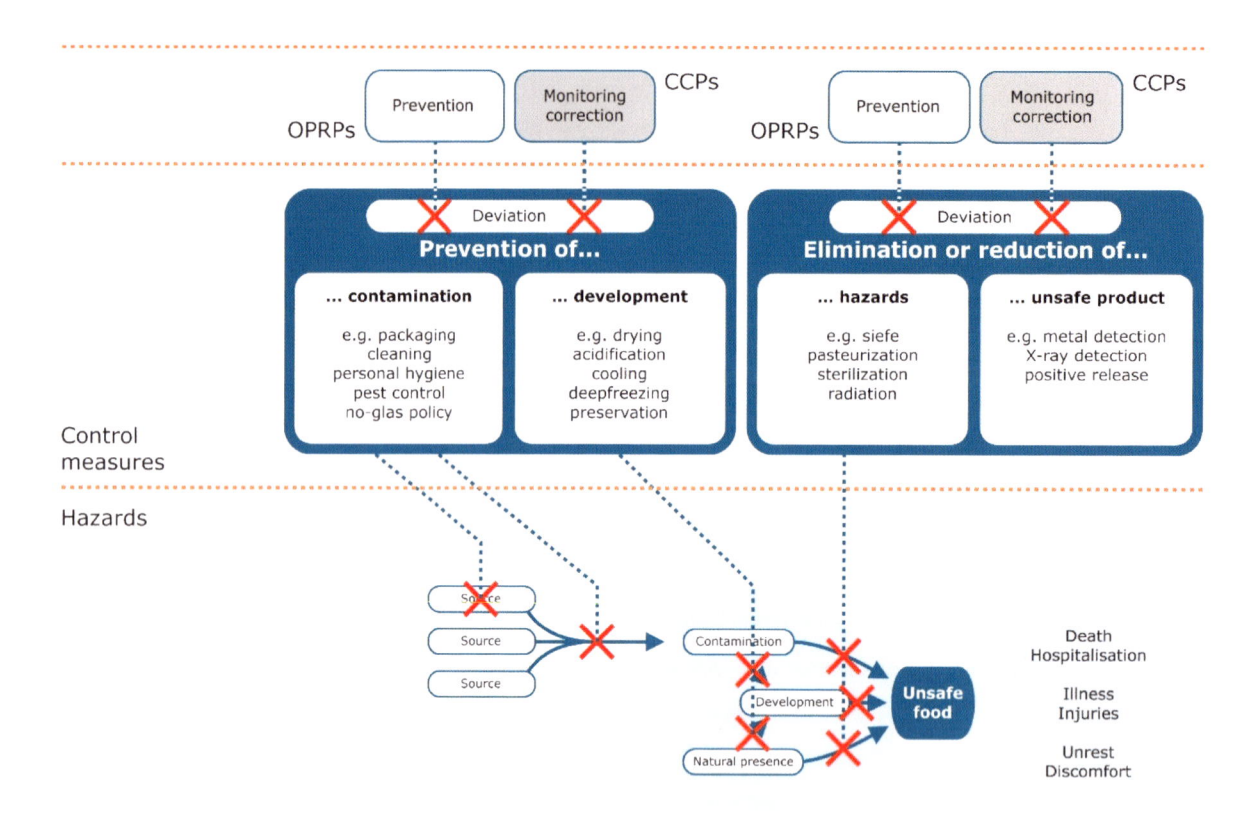

Figure 4.5 Food Safety Bow Tie including prevention and correction of deviations

4.6 Determination and validation of critical limits

In step 8 of the HACCP guidelines, the third principle is applied: determination of critical limits for each CCP. Critical limits for CCPs should be chosen in such a way that when a CCP meets the critical limit, the presence of the hazard is restricted below an established acceptable level. The HACCP guidelines state that critical limits must be measurable and this means that they refer to objective, measurable parameters such as temperature, time, pH, Aw-value or chloride concentration. In scientific literature, a lot of information can be found regarding control measures and associated critical limits, especially for microbiological hazards. For using such information, it is important that it is applicable to and consistent with the specific processes and products subject to the hazard analysis. It can be useful to establish alarm limits in addition to critical limits. Alarm limits make it possible to intervene before critical limits are exceeded.

Next to critical limits being measurable, according to the HACCP guidelines, the critical limits must also be validated. The HACCP guidelines offer no further explanation on the concept of validation at this step (step 8) but is discussed along with verification at step 11. However, validation is closely related to the determination of critical limits. This is put forward in a document called, *'Guidelines for the validation of food safety control measures'* of the Codex Alimentarius as a supplement to the *General Principles of Food Hygiene*. This document gives more specific definition of validation:

Definitions
Validation (according to Guidelines for the validation of food safety control measures): obtaining evidence that a control measure or combination of control measures, if properly implemented, is capable of controlling the hazard to a specified outcome.

Validation (according to the General Principles of Food Hygiene): obtaining evidence that the elements of the HACCP system are effective.

Verification (according to the General Principles of Food Hygiene): the application of methods, procedures, tests or other evaluations, in addition to monitoring, to determine compliance with the HACCP plan.

End of example

Validation is thus aimed at control measures and thereby is directly related to the determination of critical limits. For example, for pasteurization the choice for a critical limit at 72°C (for 14 seconds) or at 78°C, depends upon the purpose – or as stated in the definition, the 'specified outcome' – that should be achieved with pasteurization.

Typical examples of validation questions are:
- Does this disinfecting fluid with this concentration of chlorine, ensures a sufficient reduction of *Listeria* on this surface?
- How long can shelf life be at storage temperature of 7°C in order to restrict growth of Bacillus cereus below acceptable levels?

Validation, in practice, is often confused with verification. In this book validation is will be related to the design or capability of a control measure, while verification is aimed at its practical implementation or its reliability. This is consistent with the generally accepted view that validation is preferably performed prior to the implementation of a control measure, while verification is done after the implementation. Paragraph 4.8 gives additional information regarding validation.

Validation starts with the definition of an intended specific outcome or result of a control measure. For this, wherever possible, standards must be adopted, which are incorporated in legislations such as European Regulation (EC) 2073/2005 and Regulation (EC) 1881/2006. Along with food safety criteria for end products, Regulation (EC) 2073/2005 also included standards for process steps, the so called 'process hygiene criteria'. A process hygiene criterion for a certain point in a process specifies the acceptability of a product based on presence or absence of microorganisms. Validation of a process criterion should prove that the selected control measures, maybe in combination with other control measures, is capable of meeting these process criteria. The following table provides an example of process hygiene criteria from Regulation (EC) 2073/2005.

Example
"Process hygiene criteria in relation with validation"
Table 4.3 Process hygiene criteria from Regulation (EG) 2073/2005

Food category	Micro-organisms	Sampling plan		Limits		Reference analysis method	Stage where the citerion applies	Measures in case of unsatisfactory results
		n	c	m	M			
2.2.1 Pasteurized milk and other pasteurized liquid dairy products	Entero-bacteriaceae	5	2	< 1/ml	5/ml	ISO 21528-1	End of production process	Monitoring the effectiveness of heat treatment and prevention of recontamination and quality of raw materials

The column 'Measures in case of unsatisfactory results' states that 'monitoring the effectiveness of heat treatment' should be executed. This 'monitoring the effectiveness' may include two things: 1) a verification which examines if heat treatment is applied as prescribed and if potential deviations are accurately corrected, 2) an assessment of the ability of the heat treatment to achieve the desired reduction of Enterobacteriaceae. The latter is an example of validation.
End of example

Validation is preferably done by means of an objective test. In relation to microbiological criteria in Regulation (EC) 2073/2005, a challenge tests can be performed. In a challenge test, products are intentionally contaminated with a predetermined number of certain microorganism(s). Subsequently, it is assessed whether control measures are capable of reducing the number of microorganisms to a satisfactory level (as in pasteurisation) or prohibit or restrict the growth of microorganisms to a satisfactory level (as in refrigeration, lowering the pH or Aw- value). A challenge test can prove, for example, that there cannot be any significant growth of *Salmonella* or *Listeria* in a certain product.

Validation through a challenge or any other test is not always possible. This particularly applies to OPRPs that have no measurable critical limits. Wherever it is not feasible to validate through tests, validation can be based on historical data. Favourable results of microbiological research and absence of complaints subsequently will lead to the conclusion that control measures have apparently been effective. Evidently it is preferable to validate control measures prior to their implementation with objective technical test. Where this is not possible (or in case if measures are already implemented), there should at least be a statement regarding the expected effectiveness of control measures and the data on which they will be assessed. It should be determined which data will be collected, at what frequency and how this data will be analysed.

To finish this paragraph, an additional example is shown of a standard for the validation of equipment.

Example
"Validation of cleanability of processing equipment"
An example of validation is testing of cleanability of processing equipment as per the standards of the European Hygienic Engineering and Design Group (EHEDG). The EHEDG has developed a procedure (called 'Method for assessing the in-place cleanability of food processing equipment') in which processing equipment is contaminated as per a standard method, cleaned and then assessed for cleaning result. Equipment complying with the prescribed criteria is eligible for EHEDG certificate. Another example of an EHEDG guideline concerning validation is 'Challenge test for the evaluation of the hygienic characteristics of packing machines for liquid and semi-liquid products, 2000'. EHEDG publishes various interesting guidelines that can be ordered via the EHEDG website (www.EHEDG.org).
End of example

4.7 Monitoring and corrective action
For each CCP, a monitoring system must be established in order to detect potential deviations. Recording the results of monitoring is essential to demonstrate the proper implementation and application of the monitoring procedure. Records of results should include all planned measurements or observations related to critical limits. The method and frequency of monitoring must be determined in such a way that exceedance of critical limits is detected in a timely manner that allows for corrective actions that include the exclusion of possible unsafe products.

To establish a good monitoring system, first it should be determined which deviations can occur and their related causes. In other words, one must know the deviations to be able to detect them. This is illustrated in the following example.

Example
"Core temperature of canned mushrooms"
During sterilization of canned mushrooms, it is impossible to measure the core temperature in cans. The temperature is measured in a heating medium and through various tests it should be validated that the measurement in the heating medium is representative of the temperature in cans. However, when cans have high fill weight, this affects heat transfer and the temperature of heating medium might no longer be an exact reflection of the temperature in the can. A monitoring system which should indicate that the critical limits for temperature-time-combination have been reached, must also include monitoring of fill weight of cans, next to monitoring of temperature and time.
End of example

Furthermore, to establish the frequency of a measurement (or observation), it is important to know the nature and the causes of deviations. Some deviations, for example, a tear in a sieve, have a permanent character. On the other hand, a temperature drop in an ongoing pasteurisation process can be temporary. The frequency of measurements must be established in a way that allows detection of these temporary deviations. Deviations with a permanent character basically impose no requirements on the frequency of measurement. For the frequency of a measurement then applies a second argument regarding the application of corrective action. The frequency must be chosen in such a way that products which may be affected by a deviation and therefore possibly be unsafe, could be easily and completely excluded from the process. When a very low frequency is chosen then it might happen that potentially unsafe products have already left the company, and then it will take more effort to retrieve them.

In the application of corrective actions, ISO 22000 makes a distinction between corrections and corrective actions. Corrections include all actions which aim at mitigation of the consequences like exclusion of possible unsafe food, while corrective measures are concerned with the elimination of causes of deviations. In this book mitigation of consequences of deviations is interpreted as a corrective measure and elimination of causes as a preventive measure.

About corrections with reference to CCPs, ISO 22000 indicates that products which are produced under conditions in which critical limits have been exceeded, are potentially unsafe. These potentially unsafe products must be traced, identified and excluded from the process. With regard to OPRPs, ISO 22000 states that when an OPRP is not applied in accordance with the standard, it does not necessarily mean that the concerned products should be considered as potentially unsafe. If an OPRP is not correctly applied, risk for the safety of the products should be assessed with regard to the conditions, causes, nature and scope of the deviation. Depending upon this assessment, it must be decided if a product can be considered as safe or as potentially unsafe. Both, for CCPs and OPRPs, it is true that potentially unsafe products should be excluded from the process and must be blocked. Blocked products may be released and may only be released, provided that a reassessment has proved the products are safe. This reassessment must be based on data other than that from the monitoring system. Reassessment, for example, can be done through sampling and analysis. However, in many cases blocked products should be regarded as unsafe, and they must be reprocessed or destroyed.

To be able demonstrate the proper application of a control measure, it should be ensured that monitoring procedures are properly documented and the results are recorded. Documentation comprises of procedures, instructions and records concerning:
- critical limits for CCPs

- implementation standard for OPRPs;

- methods for measurement or observation;

- necessary equipment;

- calibration of equipment;

- frequency of measurement or observation;

- registration of results;

- assessment of results;

- corrective actions and their status;

- measures to eliminate causes of deviations.

4.8 Validation and verification

Verification is the sixth principle of HACCP and is considered in step 11 of the HACCP guidelines. The sixth principle also includes validation. Validation is primarily important in relation to control measures and critical limits as discussed in section 4.6. However, validation is more than assessment of control measures, validation also applies to the food safety management system as a whole. This section first discusses validation and then verification of food safety management system. In this book validation is regarded as an assessment of design of control measures (also refer section 4.6); verification is regarded as an assessment of implementation of control measures. Difference between validation and verification is illustrated in the following example.

Example
"Analysis of complaints; will that be verification or validation?"
Complaint analysis plays a role in both validation as well as verification. When a complaint is a consequence of improper identification of a problem and ineffectiveness of chosen control measures, then this problem is related to validation. When a complaint arises due to a shortcoming in the application of a control measure, while the corresponding standards, procedures and instructions are well established, then this problem is related to verification. The absence of complaints thus means that a system is well designed (validation) and properly implemented (verification).
End of example

4.8.1 Validation of a food safety management system

In HACCP, validation is defined as '*obtaining evidence that the elements of the food safety system are effective*'. Validation focuses on the design of the food safety management system and should demonstrate as objectively as possible that a system is capable of controlling all relevant hazards. Validation is preferably performed before a system is implemented. Validation is applicable to all individual components of hazard and risk analysis. The individual components for validation are:

- List of relevant hazards: there must be a complete list of hazards that include the description of health effects which is substantiated by reliable scientific information. In any case, all those hazards subject to legislation, should be included.

- Control measures including critical limits: validation of control measures including critical limits has been previously discussed in section 4.6.

- Monitoring procedures and corrective actions: the HACCP team must assess the potential deviations in control measures and the risk of these deviations to food safety. Monitoring systems must be designed in such a way that all high risk deviations will be detected. Corrective actions must exclude all potentially unsafe products from a process. A practical example of validation of a monitoring system is testing of an automatic temperature alarm system in a newly built cold storage room.

Validation is preferably based on technical evidence produced in an objective test, but for a food safety management system as a whole this will not be possible. An expert opinion is an alternative. This expert (or experts) can be part of the organisation or from a third party. It is important that the expert is competent and objective. For objectivity, validation is preferably done by an expert who is not a member of the HACCP team.

Once a food safety management system is validated, the validation remains valid, as long as there are no modifications that requires adaptation in the design of the system. However, in the dynamic world of food production, there is no lack of modifications. Modifications can be external or internal. External modifications include disclosure of a new hazard, new scientific insights about risks or hazards and new requirements in legislation. Example are: new hazards such as acrylamide (issue in 2002) melamine (issue in 2008) and modifications in the list of allergens as stated in European legislation. Modifications can also be internal. Example are changes in raw materials and recipes, installing new machinery, introduction of new products and

so on. The HACCP team must ensure that they are well informed about the modifications and ascertain that all relevant components of the food safety management system are adapted and validated.
Amendments of a food safety management system due to external and internal modifications makes validation a continuous activity. Therefore, a HACCP team must be constantly attentive to indications of all kinds of changes. Product and process developments should be pursued, legislation and literature should be updated, indications from the market should be considered, and scientific and technical developments should be tracked. In order to prove that the food safety management system is well adapted to these modifications, it is useful to verify procedures once a year before validation. This verification means that the individual components of a food safety management system are checked to detect whether there were any modifications and if and how the food safety management system was adopted.

4.8.2 Verification of a food safety management system

With regard to the principle that 'all measures can fail' it may be argued that even monitoring procedures and corrective actions might fail. The primary aim of verification is to ensure that monitoring procedures and corrective measures at CCP and OPRPs are applied in accordance to predetermined procedures. In fact, verification adds one more safety level to the control of hazards. The first level consists of control measures, the second level includes monitoring procedures and corrective measures; verification is the third level. These levels are illustrated in figure 4.9. Failure of monitoring procedures in the context of HACCP is certainly worthy of attention. The following example serves to illustrate this.

Example
"A loose contact during pasteurisation of milk"
A HACCP team has identified the pasteurisation of milk as a CCP. The critical limit is set at 76°C for 12 seconds. The temperature is continuously measured and recorded. However, due to a partially loosened contact in electronic temperature meter, a value which is higher than the actual temperature is indicated. The automated process control (by means of a PLC, Programmable Logic Controller) receives this incorrect high value and subsequently lowers the heat supply and so the actual temperature will drop. Finally, the actual temperature in this true practice case dropped to 65°C whilst the meter still indicated 76°C. Since the automated process control as well as the recording device were connected to this one temperature meter, everything seemed to go well. The fault was revealed only when verification based on a microbiological test showed an increase in the number of bacteria. Problematic in this case was that the microbiological test results was known rather late.
The above example is shows a deviation in a monitoring system. The cause is the loosening of a contact. A preventive measure for this cause would be a better connection of the contact of the meter, for example, by welding or soldering instead of screws. An example of a corrective measure for this deviation is the monitoring of the temperature meter by installing a second temperature meter. By comparing the value of one temperature meter continuously with the other, a deviation in the value indicated by one of these two meters can be swiftly detected, and measures can be undertaken for correcting the deviation. In fact, this means that the second meter verifies if the first meter indicates a correct value.
In some branches of the industry, it is a good practice to have a duplex measurement system. One measurement system is used for direct control of the process, the other system is used to monitor the process and gather information for the control of quality and safety.
End of example

One way to control the problem of failure of instruments is through calibration. The frequently of calibration must be sufficient to allow timely corrective action. When major deviations suddenly occur in instruments that are used to monitor a CCP, the usual yearly calibration has little contribution. Calibration with a frequency of once per year has value only to relatively slow degradation of accuracy of measuring instruments.

Another example of the failure of a monitoring procedure is the situation in which employees forget to monitor a CCP. Preventive measures, for example the training of employees, can be used to reduce the probability of such a deviation. The probability of omission of a monitoring procedure is also reduced when a measurement is easy without much effort. Based on this kind of thinking, daily verification of monitoring procedures at a CCP is a sensible thing to do.

For most components of a food safety management system, one verification per year is a standard frequency. For components like OPRPs and CCPs a higher frequency should apply. It is highly recommended that companies on a daily or weekly bases verify the procedures, instructions and registrations related to CCPs and OPRPs. A unfavourable result in the verification of a CCP indicates that the control at that moment was not guaranteed and consequently there may be unsafe products. Also procedures, instructions and registrations related to OPRPs must be verified with a relatively high frequency. The appropriate frequency is determined by the nature of an OPRP and the risk of potential deviation. The probability of deviations in an OPRP can strongly depend upon, for example, the discipline of employees in production; a high degree of discipline can mean that the frequency of verification can be reduced. This can apply to, for example, pest-control, assessment of good housekeeping and glass breakage.

Internal audits are an important type of verification. A food company must have procedures for planning and execution of internal audits. The choice of auditors and the way in which audits are conducted must ensure the necessary objectivity and impartiality; auditors, therefore, should not audit their own work. Next to assessment of the correct application of control measures, verification also includes an assessment of obtained results. So verification should include the evaluation of cases of non-compliant products, rejected raw materials, retrieval actions, complaints from buyers and consumers and results obtained from sample analysis.

Example
"Sampling and analysis, is that a control measure or a verification?"
In paragraph 4.3, sampling and analysis is presented as a control measure while, it can also be considered as verification. Both these forms occur in practice.
Sampling and analysis can be used as a control measure with which safe and unsafe products can be separated from each other. Or in other words, safe products are positively released. This means that the frequency of sampling and analysis must be statistically justified. This also means that sampling and analysis must be incorporated as a step in the process, as shown in fig. 4.6.

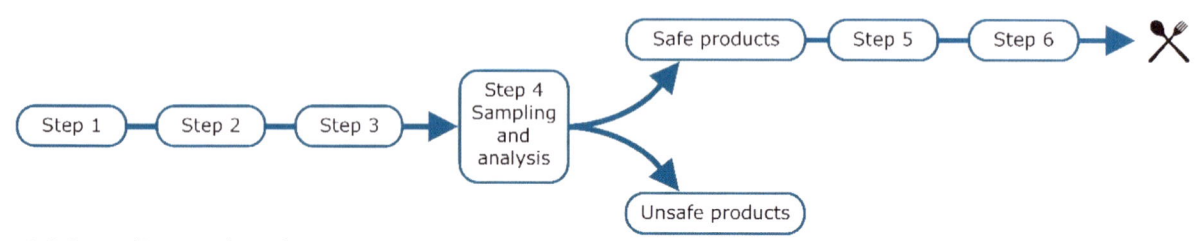

Figure 4.6 Sampling and analysis as process-step

Sampling and analysis can also be used as a type of verification, see fig.4.7. The sampling and analysis in this case means an extra check, the actual control and monitoring takes place in step 4, pasteurisation. Results provided by verification will be a confirmation of the control. There are no specific requirements regarding sampling frequency. Unfavourable results from sampling and analysis indicate that the control measures and the monitoring have failed.

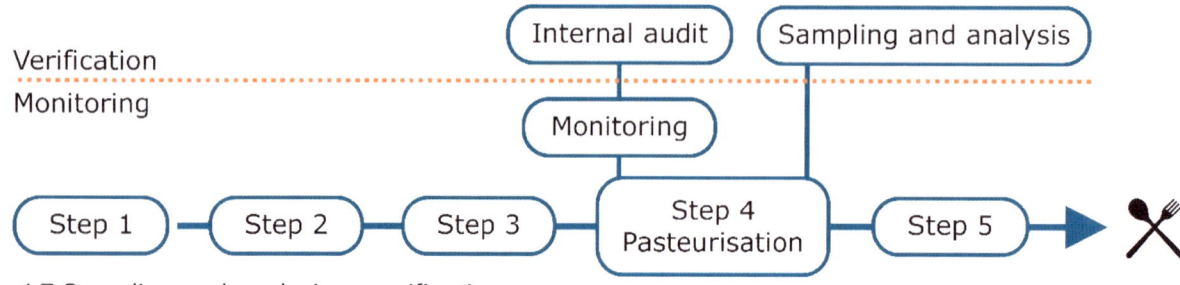

Figure 4.7 Sampling and analysis as verification

End of example

Validation and verification of a food safety management system provide important information for annual management review. A management review is not a component of the formal HACCP system as described in the *General Principles of Food Hygiene*, but is a requirement in almost all certification standards.

4.9 Documentation and registration

Food companies should not only ensure that food safety hazards are controlled but should also be able to demonstrate this control. Documentation and registration, therefore, are important and relevant to all principles and all steps of design, implementation, application, maintenance, adaptation and improvement of a food safety management system. Hence, it is important to ensure a proper management of documents right from the commencement of a HACCP project. The application of HACCP is perfectly consistent with the use of quality management systems such as ISO 9000. It is recommended to make a structure that is suitable to the structure of an organisation. In practice, often a structure based on certification standards is chosen. Perhaps this is convenient for an external auditor and a quality manager, for other parts of the organisation a food safety management system is rather a puzzle, see figure 4.8.

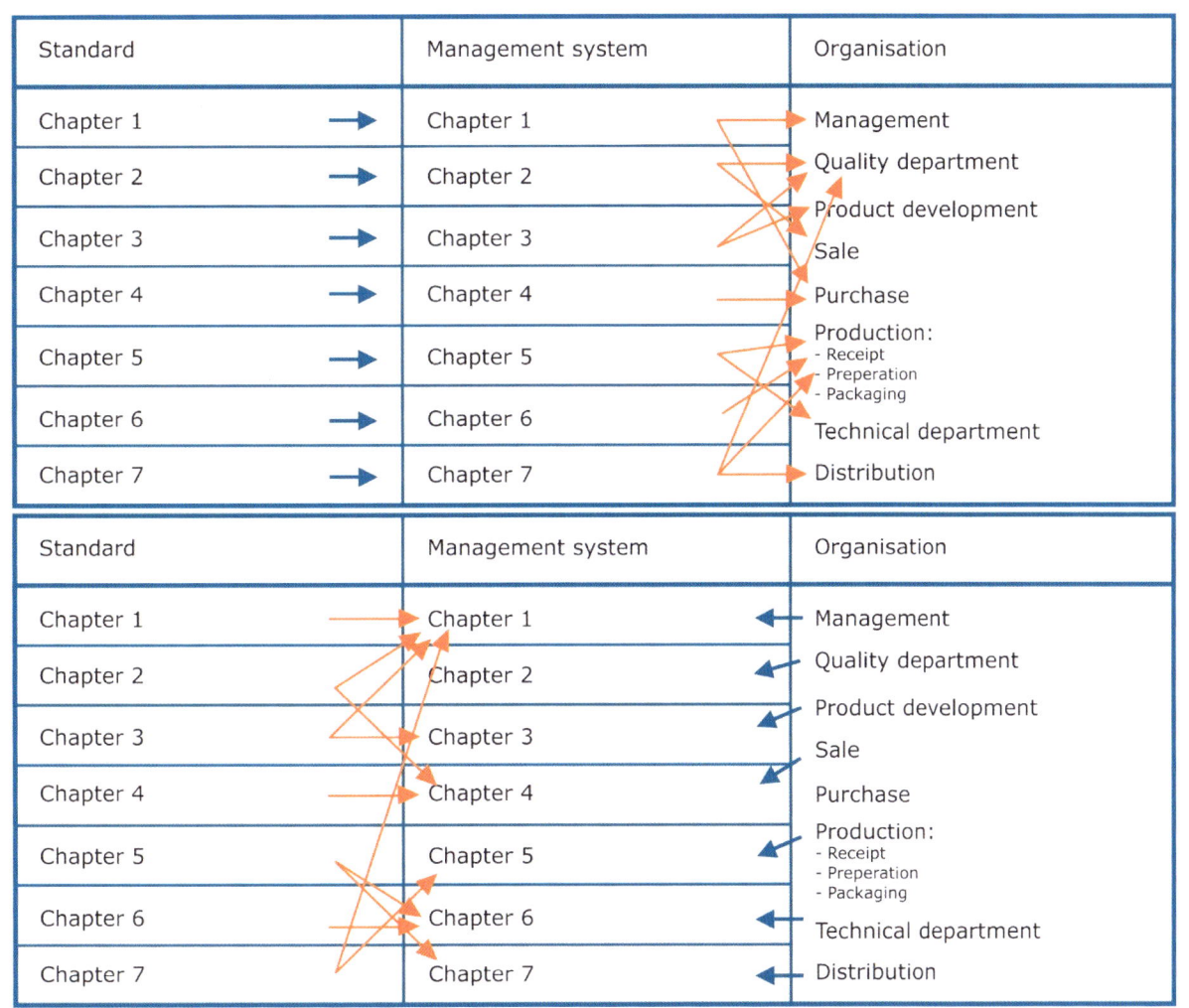

Figure 4.8 Structure of a management system (upper part: a standard based management system, lower part: an organisation oriented system)

A procedure for the management of documents must ensure that:
- documents are approved before they are issued;

- documents are updated when necessary;

- changes and the current revision status of documents are identified;

- the current version of the relevant documents are available to employees;

- documents are legible and easily comprehensible;

- obsolete documents are archived.

Furthermore, a procedure should be determined to ensure storage, protection, retrieval, retention period and destruction of records. Records must remain legible, easily comprehensible and retrievable. The available records should be concerned with:
- results of monitoring procedures;

- deviations and corrective actions;

- results of verification and internal audits;

- data regarding traceability;

- complaints and related handling.

4.10 Food Safety Bow Tie
In the preceding paragraphs, a Food Safety Bow Tie is constructed step by step on the basis of HACCP principles. A complete model is displayed in figure 4.9. To conclude this chapter, a few more comments on this model.

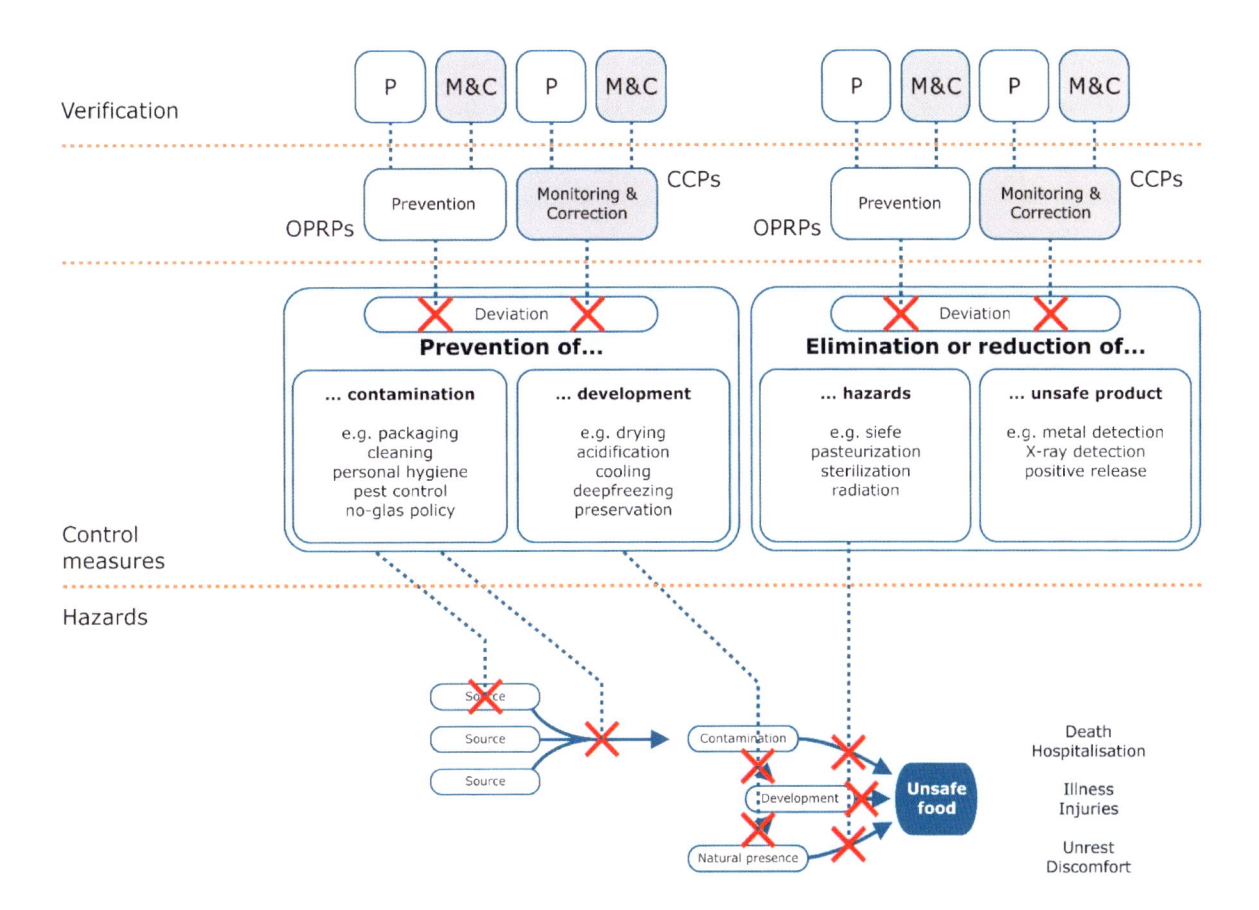

Figure 4.9 Food Safety Bow Tie

So far most attention in this book was directed to the left side of the Food Safety Bow Tie that deals with the causes of unsafe food. The model offers visibility to different safety levels, which are important for the control of food safety. The primary level contains the control measures, monitoring is at the secondary level, the tertiary level consists of verification activities. The right side of the diagram deals with measures that, once a product is in the market, could mitigate the consequences. The following text discusses a couple of measures that food companies can take in relation to unsafe products on the market.

Warning on the label
When no, or only insufficient, control measures can be undertaken, a food company should ensure that consumers are aware of potential hazards in a product. When there is uncertainty regarding the presence of this awareness, the company should make sure that consumers are informed. This can be achieved by means of a specification or a label. For example, allergens of natural presence in food cannot be prevented, eliminated or reduced. As consumers are often unaware of ingredients present in food, a declaration of all ingredients with allergens is an appropriate way to deal with this risk. European legislation contains a legal obligation to do so. Products with possible traces of allergens that are foreign to the product, can be labelled with a warning such as 'may contain traces of nuts and peanuts'. For such warnings there are no legal obligations and the company has to make a decision. Another example of such a voluntary warning is an indication on the label of honey not to use the product as food for children under 1 year of age. This is associated with infant botulism.

Recall (not included in the Food Safety Bow Tie)
Second example of a measure on the right section of the Food Safety Bow Tie is a recall. A recall is started after detection of a problem on the market. Consequently, the concerned batch of product is blocked and retrieved from the market. Good traceability is thus very important. Various certification standards include requirements for testing of recall procedure. Through the test, possible failure of a recall procedure can be detected and actions for improvement can be taken.

Chapter 5: Risk matrix and decision tree.

Chapter 4 discussed the principles and definitions of HACCP and visualized them in a Food Safety Bow Tie. This chapter deals with two important tools for hazard and risk analysis: risk matrix and decision tree.

5.1 Risk matrix

The assessment of food safety risks is done at several levels. At government level, risk assessment is considered under European Regulation (EC) 178/2002. This is a scientific process that falls under the responsibility of EFSA. Companies executing hazard and risk analysis, cannot be expected to do that in a scientifically justified way since risk assessment is too complex. However, it may be expected from the companies that they are acquainted with the results of scientific risk assessment and have a clear picture of their contribution to risk control. For food companies, a qualitative risk assessment is sufficient to recognize the importance of various control measures. Risk assessment within the context of HACCP has originated from indications in the HACCP guidelines that specify that wherever possible, the probability of occurrence of hazards and the severity of adverse health effects should be included in the hazard analysis. Risk matrices are widely used for the practical assessment of the risk of hazards. This book uses a risk matrix that is shown in figure 5.1. The matrix is partly based on the FAO matrix, see section 2.5 - figure 2.4.

Severity description	Severity	Code	NR	PI	S	M	G
Death, Disability, Hospitilazation	Very serious	VS	Very low	Low	Medium	High	High
Illness with absenteeism, Injuries, Medical treatment	Serious	S	Very low	Very low	Low	Medium	High
Discomfort, unrest	Less serious	LS	Very low	Very low	Very low	Low	Medium
No effect on consumers health	No effect	NE	Very low	Very low	Very low	Very low	Low
			Not relevant	Practical impossibility	Small	Medium	Big
			No consumers affected	Less than 1 consumer affected in 100 years	2 or less consumers affected in 50 years	2 to 50 consumers affected in 50 years	More than 1 consumer affected in 1 year

Figure 5.1: Risk matrix for food safety

Significance of different risk levels is as follows:
- High risks: risk is unacceptable. Products are deemed unsafe and shall not be presented for consumption.

- Medium risks: the risk is acceptable on certain conditions. Products are regarded as safe when provided the specified conditions are fulfilled. The conditions could include that the risk is communicated to users and/or consumers and that there is an ongoing effort to gain further reduction of the risk.

- Low risks: risk is assumed as acceptable. Products are generally considered to have a low risk and can be regarded as suitable for consumption. The argumentation of low risk must be incorporated in hazard and risk analysis. There should be an indication regarding measures that contribute to this low risk.

- Very low risks: risk is deemed to be acceptable. Products are considered to have a very low risk, therefore, suitable for consumption.

No known scientific research has been done on the role of a risk matrix related to food safety. This means that the risk matrix above is not validated. The first possibility for validation would be to align the matrix with the existing accepted concepts concerning food safety. An example of such a concept could be an *Appropriate Level of Protection* (ALOP), discussed in section 2.2. Another example of a similar concept is the *botulinum-cook* for sterilized products. With this heat treatment, a twelvefold decimal reduction (12 D) of *Clostridium botulinum* is achieved. In general, this type of sterilization is considered sufficient to ensure the safety of long shelf-life of canned products.

Another possibility to attain a validated risk matrix is by designing a matrix on the basis of a current food safety situation. The current situation is then considered as acceptable but must be improved. The matrix in any case should not permit deterioration with respect to the current situation. Such a matrix could be based on the food safety situation in a country, in a certain sector of a food chain or even a situation in a certain company or in a certain process. This approach would represent different matrices for different sectors in a food chain and for different companies or different processes. This is, for the time being, fetched too far. Validation of a risk matrix requires a quantitative approach and thus necessitates a lot of reliable data. These data do not or barely exist, and it is doubtful if it will ever be there. A risk matrix based on the formula, risk = probability × severity, so far is no more than a simple qualitative tool. It helps companies to understand the risks and compare them; however, the matrix offers no quantitative base to make decisions.

The following observations must be made regarding risk matrix in figure 5.1. The probability is expressed as the number of affected consumers in a certain period. This is a quantitative approach assuming an accuracy which factually does not exist due to a lack of data. Because of the educational value, it was decided to stick to this type of expression; it forces companies or a HACCP team to think about the target group, consumers and their health and this helps to compare the probability of different risks.

5.1.1 Severity of hazards

The HACCP team has made a list of hazards, that are relevant to the products of the company. This list focuses on the left section of figure 4.1. The right section of this figure shows the possible consequences of these hazards. These consequences or the severity of hazards are the subject of this paragraph.

The severity of hazards in this book is described on the basis of the following statement:
food companies have no influence on the severity of hazards; they only have influence on the probability.

This statement implies that the elements, severity and probability in the formula risk = probability × severity, can be separated and cab be assessed independently. A HACCP team will estimate the factor severity on the basis of literature research. The factor probability will be determined in the hazard and risk analysis to assess the need to implement control measures. The statement primarily means that the severity of a hazard should be the same for each and every product, process or company. For example, pieces of glass in the products of company 'A' are as severe as pieces of glass in the products of company 'B'. However, a few remarks must be made.

In real life, probability and severity cannot be separated; they are highly interconnected. To illustrate this: it cannot be stated that the effect of *Salmonella* is 'more severe' than the effect of a foreign object such as a stone. In a 'worst-case scenario', both a Salmonella and a stone can lead to death of a consumer. What could be claimed is that the probability of death due to consumption *Salmonella* is higher than the probability of death after ingesting a stone.

The factor probability can be evened out when the effect of two hazards is compared on the basis of an equal number of actual cases. When 1000 historical cases of consumers who suffered a *Salmonella* infection are compared with 1000 historical cases of consumers injured due to a stone in their food, then it becomes clear that the total loss of good health caused by 1000 *Salmonella* infections would be considerably higher. So on average, the effect of *Salmonella* is more serious than that of a stone. This is the concept that forms the base of this paragraph.

An independent assessment of the factors probability and severity leads to the idea that one hazard is more serious than the other. In fact, this contradicts the adage of Paracelsus (1493-1541): 'All substances are poisonous; there is none, that is not a poison. The right dose differentiates a poison and a remedy.' This is particularly true for chemical hazards. For physical hazards and allergens the dose of the hazards is less relevant. For microbiological hazards the dose response relation is rather unclear. Some bacteria have great capability of surviving acid stomach and good capability to attach to the tissue of intestines and then cause infection. For bacteria that produce toxins, number up to 100.000 per gram in food are needed to cause any effect. Possible health injury resulting from a hazard, however, also depends on the condition of the consumer, see the example - influence of the condition of the consumer. For companies dealing with the vulnerable groups, like infant food formula, this means that the severity of hazards must be estimated higher than the companies producing food for general population. Finally, the medical treatment of an adverse health effect is a significant determinant of the total loss of good health.

The statement that the companies have no influence on the severity of hazards entails a necessary nuance in some cases. However, for the sake of simplicity, the statement is valid and concept is that the severity of hazards should be in the same category in every hazard and risk analysis.

Example
"Influence of the condition of the consumer"
In 2001 there was an outbreak of Salmonella enteritidis Pt 6 epidemic in a nursing home in Zwolle, the Netherlands. Based on complaints such as diarrhoea and fever, 63 residents, 40 staff members and 32 patients were suspected of incurring Salmonella infection. For 70% of the residents Salmonella infection was confirmed and among patients and staff these were 47% and 25% respectively. Finally, 46 people fell sick and 5 people succumbed due to this food infection. An uneven distribution of the percentage of Salmonella positive cases among patients, residents and staff members indicates that not everyone is evenly susceptible to the bacteria. Elderly people and patients have a less effective immune system. Furthermore, low acidification rate of the gastrointestinal contents of elderly people greatly influences the likelihood of the survival of Salmonella in the stomach, see figure 5.2. The probable source of the contamination was a desert with raw egg.
End of example

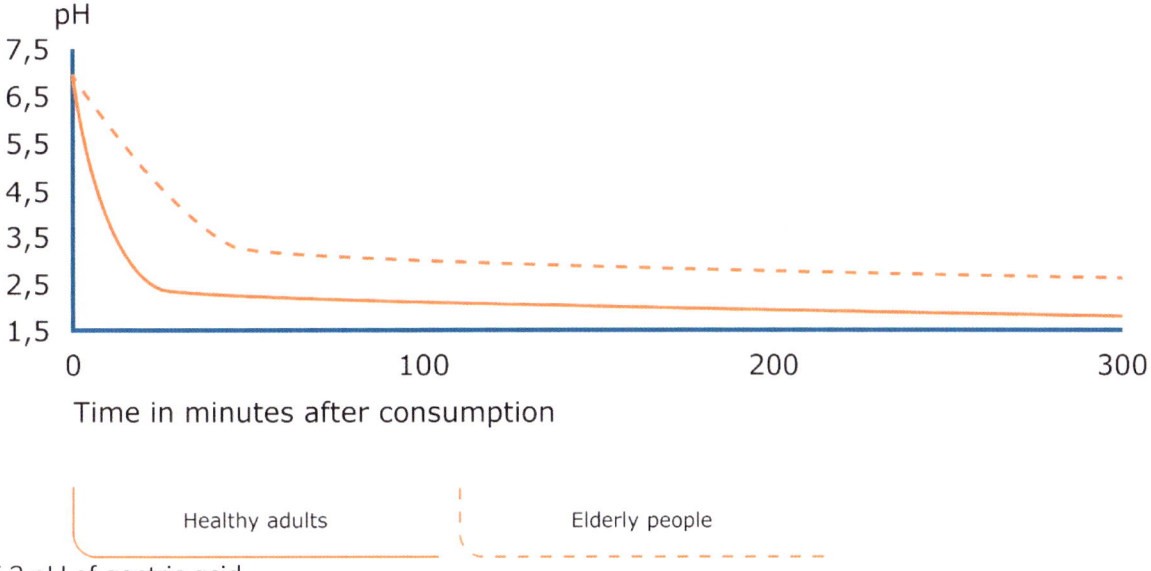

Figure 5.2 pH of gastric acid

For good understanding of the risks of unsafe food, determination of the severity of health effects of different hazards is a key element. This section attempts to categorize hazards according to their severity. The categories are related to the categories in the risk matrix. Making such a list is a cumbersome issue for a HACCP team. Research of health effects of hazards lies within the scope of medical microbiologists and toxicologists. Large number of expert publications are available and over the past years the amount of reliable and usable information has increased enormously. Fortunately, there are also some statistics available, which can substantiate an assessment of severity of hazards.

One way to statistically compare the severity of various types of hazards with each other is by expressing the burden of disease in DALY's (Disability Adjusted Life Years). The burden of disease is regarded as the total time a group of people had to spent on being sick. The concept originates from the Global Burden of Disease-study (GBD) of the World Bank and the WHO. The DALY is constructed from two components: the years lost due to premature mortality and the years lived with sickness. These years with sickness are 'weighted' for the severity of disease with the help of weighting factors so that they can be compared with the life years which are lost due to mortality. See the example *Disability Adjusted Life Years*.

Example
"Disability Adjusted Life Years – An example"
The disease burden in DALY's is calculated for a group of three fictitious persons. Life expectancy for all is set at 80 years.

The first person is someone who meets with a fatal car accident at the age of 40. With respect to the life expectancy of 80 years, this person has lost 40 life years and thus 40 DALY's.

The second person gets rheumatism at the age of 50. The weighting factor for rheumatism is set at 0.5. This means that the number of years to live with rheumatism is considered as 'lost' for half of it. The number of years to live with rheumatism is 30 (the life expectancy of 80 minus 50). This person thus lost 0.5 × 30 = 15 disease year equivalents. He does not die earlier and consequently, loses 15 DALYs.

The third person gets diabetes mellitus at the age of 30 (weighting factor 0.2) and dies at the age of 60. This person lost 20 life years with respect to the life expectancy of 80 years. He has lived with the disease with weighting factor 0.2 for 30 years: this is 30 × 0.2 = 6 disease year equivalents. Together 26 DALY's.
End of example

In chapter 4 the report 'Ons eten gemeten (RIVM – 2004) gives an overview regarding the disease burden due to hazards in food. According to this report, the total disease burden of food infections amounts between 1,000 and 4,000 DALY's per year. This represent 300,000 to 750,000 cases of sickness (gastroenteritis), around 25,000 physician visits and 20 to 200 deaths. For chemical hazards, including allergens, the total health loss is estimated to approximately 2,000 DALY's per year, however, the uncertainty is large. About physical hazards, so far no data has been published regarding the total disease burden. The following figures, as calculated for diseases in the Netherlands, are presented to provide a general idea of the meaning of DALY figures and the magnitude of disease burden. In the assessment of the year 2000, the coronary heart diseases top the rank list with approximately 350,000 DALY's. Other examples are of rheumatoid arthritis with about 70,000, schizophrenia with approximately 20,000 and urinary tract infection with around 10,000 DALY's.

Severity of microbiological hazards
The difference between severities of microbiological hazards is clearly indicated in table 5.1. The table shows different microbiological hazards arranged in order of decreasing severity. In this table, it should be noted that there is a certain degree of uncertainty regarding the actual number of DALY's. The absolute level of the numbers is therefore, strongly dependent on the chosen statistical reliability. The degree of uncertainty for all microorganisms mentioned in the table is equal; which means that order decreasing severity is still reliable.

Table 5.1 Disease burden (source: RIVM – Disease burden and costs of selected food-borne pathogens in the Netherlands, 2006)

Pathogen organism		Total disease burden in the Netherlands in 2006	Disease burden per incident
		DALY	DALY per incident
Listeria monocytogenes (perinatal)	bacteria	237	47.4
Toxoplasma gondii (congenital)	protozoa	620	5.64
Listeria monocytogenes	bacteria	108	2.25
Hepatitis E virus	virus	136	0.450
Hepatitis A virus	virus	103	0.080
E. coli STEC 0157	bacteria	117	0.063
Salmonella sp.	bacteria	1,053	0.024
Campylobacter sp.	bacteria	1,833	0.023
Clostridium perfringens	bacteria	509	0.003
Rotavirus	virus	841	0.003
Staphylococcus aureus	bacteria	688	0.002
Bacillus cereus	bacteria	109	0.002
Norovirus	virus	1.083	0.002
Cryptosporidium parvum	protozoa	77	0.001
Giardia lamblia	protozoa	160	0.001

Note: both perinatal and congenital are related to an infection arising in mother and subsequently transmitted to the foetus.

Furthermore, table 5.2 gives an excellent insight into the relative severity of different microbiological hazards.

Table 5.2 Ratios of hospitalizations and deaths (source: Food-related Illness and Death in the United States, September-October 1999)

	Total cases of illness	Total hospitalizations per 1000 cases of illness	Total deaths per 1000 cases of illness

Listeria	2,518	930	200
Botulism	58	800	80
E. coli 0157:H7	73,480	300	9
Salmonella	1,412,498	220	7
Campylobacter	2,453,926	100	1
Bacillus cereus	27,360	6	0.1

Severity of physical hazards

Assessment of total disease burden in DALY's is not available for physical hazards. However, an article published in 2000 has cited a good example of the severity of physical hazards. The article provides an insight into the work done by Health Hazard Evaluation Board (HHEB), a commission which is a part of the Food and Drug Administration (FDA, United States). The commission evaluates the incidents of food hazards and gives its opinion on their risks.

The commission uses of a set of four categories to classify health effects ranging from 'no' via 'low' and 'moderate to severe' to 'life-threatening' and thereby also considers the period wherein the adverse health effects are experienced. In the years 1972 to 1997, the commission has given its opinion on more than 4000 cases. 190 among these cases involved foreign objects in food products. The commission concluded that none of the incidents concerning physical hazards was classified in the life-threatening category. 29 cases were classified in moderate category; 119 cases were considered as low and 42 cases were without any risk. See table 5.3. In the right column of the table, an extrapolation of the research data of cases from 190 to 1000 is shown. It cannot be ruled out that a physical contamination can lead to death (suffocation due to a fish bone in the throat is 'quite common'); hence '1' is filled in extrapolation in the life-threatening category. The table gives no distinction between various hazards in the category of physical hazards. It is plausible that the total loss of good health due to thousand incidents pieces of glass is higher than the loss of good health due to thousand incidents of stones.

Table 5.3 Severity of physical hazards

	Research results	Extrapolation to 1000 cases
Life threatening	0	1
Moderate to severe	29	153
Low	119	626
No	42	222
Total	**190**	**1000**

An overview of the Dutch Food Safety Authority about consumers' complaints concerning physical hazards in 2006 and 2007 shows that 28.2% of the complaints were about glass, then followed by metal fragments with 27.4% and plastic pieces holding 23.6%. The remaining 20% included objects such as pieces of bones, stones and wood (source: VMT, 2009, nr.5).

Severity of chemical hazards

Regarding chemical hazards in food products, the actual occurrence of perceptible health effects in many cases is uncertain or even unknown. In fact, this is proved only for a limited number of substances producing acute effects such as some natural toxins. Reliable and well substantiated figures of disease burden of chemical hazards are hardly available.

In the absence of statistics, the severity of chemical hazards can be assessed by means of symptoms and working mechanism. The category of chemical hazards includes a large number of diverse hazardous substances varying greatly in their mechanism and health effects. This makes the mutual comparison and ranking of the severity of these hazards very difficult. Besides, health effects of chemical hazards frequently depend on the total intake, where as some substances can accumulate in the body. Furthermore, it is possible that these substances, to a smaller or greater extent, come from sources other than food, like for example polluted air, medicines or personal-care products.

Comparing and ranking the severity of chemical hazards is beyond the scope of this book. This subject demands an expertise and a diligence which falls neither within the context of this book nor within the competency of the author. The following example concerning carcinogenic substances is thus merely an illustration.

Example
"Carcinogenic substances"
Carcinogenic (cancer causing) substances are classified into two categories based on their working mechanism: genotoxic and non-genotoxic. For genotoxic carcinogens, it is assumed that at every level of exposure, there is a chance of cancer. For these substances there are no thresholds below which absolutely no hazardous effects are observed. For non-genotoxic carcinogens, an existence of a threshold is assumed. No adverse health effects appear below this threshold level of exposure. It is reasonable to consider the genotoxic carcinogens as 'more severe' than the non-genotoxic carcinogens. An example of a genotoxic carcinogen is aflatoxin. The particular characteristic of aflatoxin is its ability to cause liver tumours. (Source: EFSA http://ec.europa.eu/food/fs/sc/scf/reports/scf_reports_35.pdf.)

Also with regard to the toxicity of pesticides, the level of exposure plays an important role. In legislative procedures for authorized use of a pesticide it is considered that for some pesticides even a single high dose of residues can lead to health damage. In such a case, an Acute Reference Dose (ARFD, expressed in mg per kg of body weight per day) is determined. Excess of ARFD poses a direct risk to health. ARFD has been determined for around 150 substances.
End of example

Severity of allergens
For allergens as well there are only limited data regarding the number and the nature of diverse allergic reactions and the total disease burden. It is estimated that in the total population of the Netherlands approximately 1-3% of adults and 4-6% of children have food allergy or food intolerance. In absence of statistics, the severity of allergens should be estimated on the basis of symptoms and the working mechanism.

The exposure of human beings to proteins foreign to the body generally leads to a reaction of the immune system. Normally, the immune system does not react to foreign proteins that are ingested via food (oral tolerance), but for some individuals certain proteins lead to immunological reactions. This is known as food allergy. The symptoms of food allergy vary from mild to life threatening. The mildest symptoms are, for example, itching, nasal congestion, hoarseness, nausea and stomach ache. Some more severe symptoms are eczema, oedema and diarrhoea. Allergens are life threatening if they induce an anaphylactic shock. Next to this, some non-immunological reactions to food constituents may occur. This is then called a food intolerance. Sometimes the symptoms of food intolerance resemble to the symptoms of allergy, but they are mostly mild and never life threatening. Coeliac disease (gluten intolerance) is an intolerance that should be considered as serious.

Allergic reactions greatly differ among individuals which makes it difficult to categorize the severity of different allergens. Furthermore, due to these individual differences, it is very difficult to determine threshold values below which no effects can occur. For eggs, it has been reported that the consumption of 10-190 mg of egg albumin can cause serious allergic reactions which require medical treatment. In a research concerning cows' milk, allergic reactions were identified after exposure to proteins ranging from 1 microgram to 6 grams. A fatal incidence was reported after exposure to 60 mg of casein. Regarding peanut allergy, the amount of allergens required for a person to present symptoms varies from 0.1 to 50 mg. In general, it is believed that peanuts and nuts mostly lead to very serious reactions. Quantitative substantiation of this is limited.

Severity of hazards in the risk matrix

In a nutshell, it can be said that comparing the severity of individual hazards is not easy. Microorganisms are rather easy to compare especially due to the availability of a relatively large amount of data. For physical hazards, limited data are available. This data supports the obvious idea that the severity of physical hazards should be assessed lower than the severity of most microbiological hazards. For chemical hazards and allergens, there is insufficient reliable data concerning disease burden or severity of consequences. Therefore, for chemical hazards and allergens, the severity should be estimated on the basis of the mechanism and symptoms. Genotoxic carcinogenic mycotoxins such as aflatoxin should be assessed as very serious just as the allergens that can cause an anaphylactic shock.

Figure 5.3 illustrates the use of a risk matrix to estimate the severity of a hazard. Two out of the three hazards in figure 5.3, *Listeria monocytogenes* and *Salmonella sp.*, have a relatively high degree of severity and also are fairly common. Hence, both hazards should be considered in any food safety management system. *Listeria monocytogenes* and/or *Salmonella sp.* can be designated as 'non-relevant' in the list of relevant hazards; this can only be accepted when these hazards have never led to any incident in the products of a company and also not in the similar products in the sector and when such incidents can be ruled out in future as well. This last part must be assessed from the situation in which all potential control measures would be omitted. Whenever *Listeria monocytogenes* and *Salmonella sp.* are regarded as non-relevant in a hazard inventory, an argumentation must be incorporated. When *Listeria monocytogenes* or *Salmonella sp.* are regarded as relevant, the hazard and risk analysis should points out where these hazards could occur, and include the associated risk and the applicable control measures.

The list of relevant hazards is completed when the severity of hazards is determined. An important part of the formula risk = probability × severity has now been established. The probability of occurrence of the adverse health effects can be reduced with control measures. This is the subject of the following section.

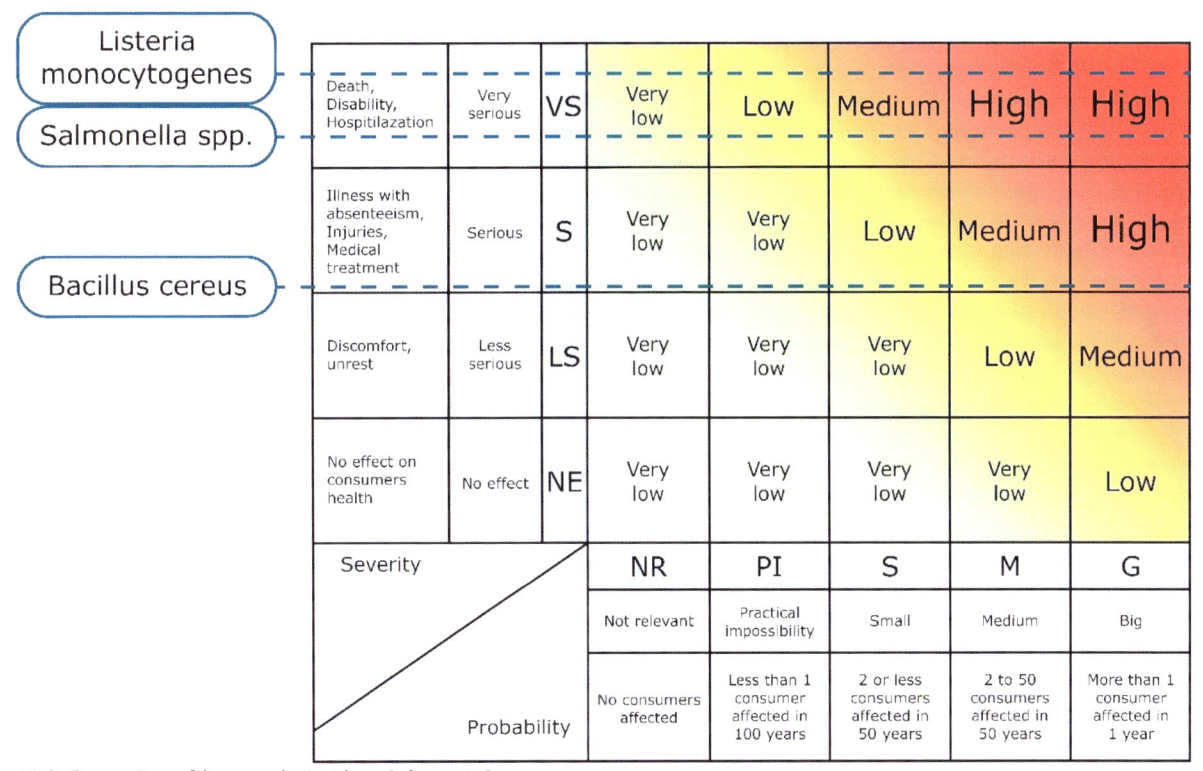

Hazards	Severity description			NR	PI	S	M	G
Listeria monocytogenes / Salmonella spp.	Death, Disability, Hospitilazation	Very serious	VS	Very low	Low	Medium	High	High
	Illness with absenteeism, Injuries, Medical treatment	Serious	S	Very low	Very low	Low	Medium	High
Bacillus cereus	Discomfort, unrest	Less serious	LS	Very low	Very low	Very low	Low	Medium
	No effect on consumers health	No effect	NE	Very low	Very low	Very low	Very low	Low
Severity				NR	PI	S	M	G
				Not relevant	Practical impossibility	Small	Medium	Big
			Probability	No consumers affected	Less than 1 consumer affected in 100 years	2 or less consumers affected in 50 years	2 to 50 consumers affected in 50 years	More than 1 consumer affected in 1 year

Figure 5.3 Severity of hazards in the risk matrix

5.1.2 Probability of hazards

Assessment of probability of occurrence of hazards is primarily done in a situation in which there are no control measures. This situation is described in figure 5.4. Hence, it is important to know the sources of contamination and the possibilities of growth, at every process step. The higher the pressure of contamination and/or the more the possibilities of growth, the higher the probability of a hazard and of its adverse health effect.

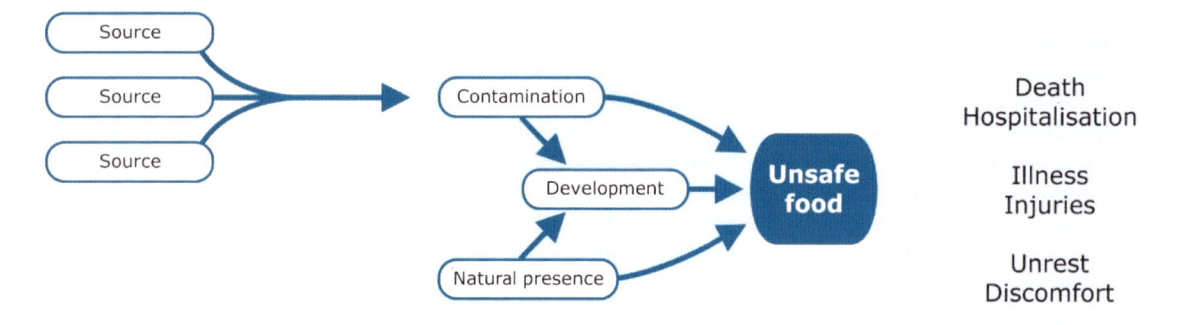

Figure 5.4 Basic Food Safety Bow Tie.

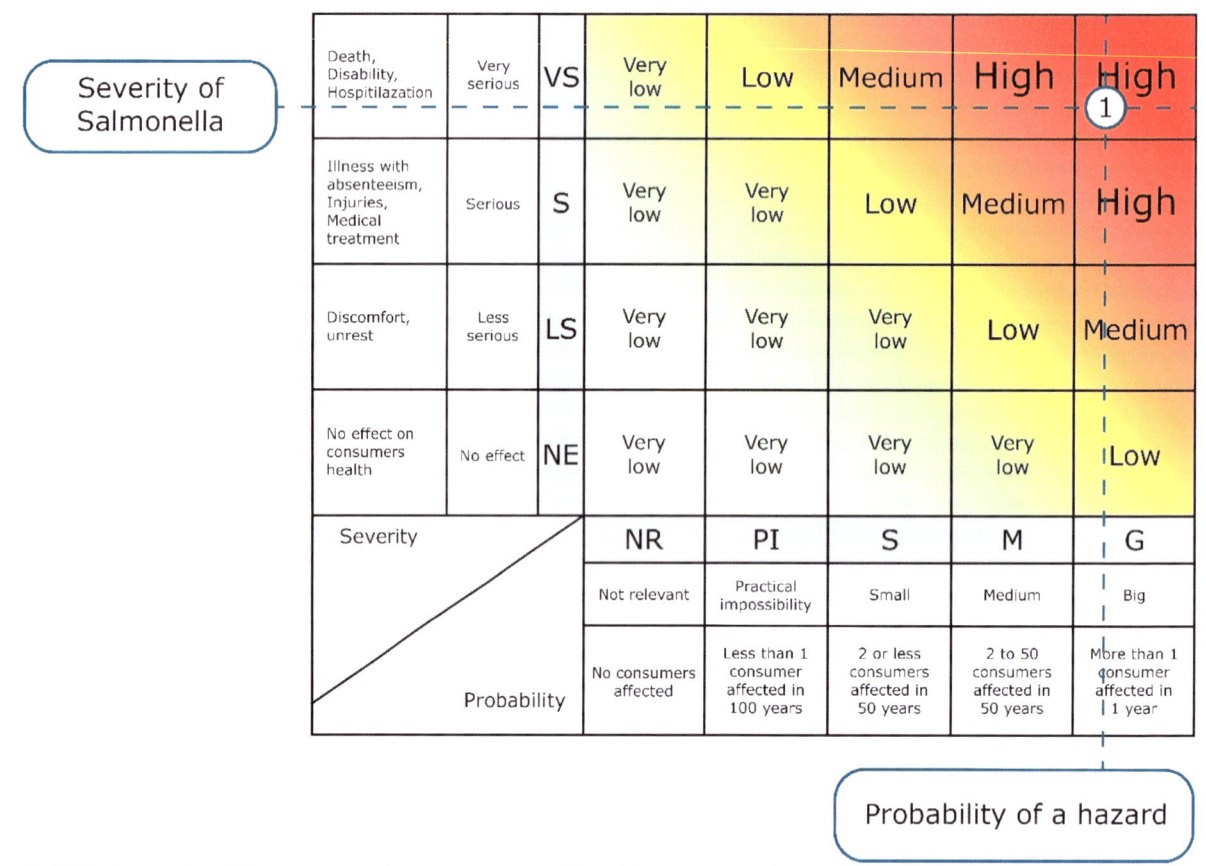

Figure 5.5 Risk matrix without control measures. A risk without control measures is indicated in point 1 of this figure.

To fully understand the relevance of a control measures and to be able to assess the risk of its failure, it is necessary to assess the risk in the situation in which the control would not exist. This sometimes leads to somewhat funny ways of thinking. For control measures that eliminate or reduce hazards such as a sieve or a metal detector, it can be easy to imagine the situation in which the control measure would not be present. For control measures that prevent contamination, like personal hygiene, cleaning and sanitation or packaging, it is not always easy to image what the situation would look like when these control measures would not be there.

When in a situation without a control measure, the risk of a hazard is estimated as high or medium, control measures are necessary. The control measures must ensure that the probability of a hazard is reduced and results in low risk. Control measures, that eliminate or reduce hazards are often capable of reducing a high risk to low risk in one go; see the arrow from point 1 to point 2 in figure 5.6. Control measures that prevent contamination and/or growth, often consist of a combination of measures, where one measure might have a greater contribution than the other (see three extended curved arrows from 1 to 2 in figure 5.6). Control measures that eliminate or reduce a hazard, can also be implemented in a subsequent step; then there are no control measures necessary for the preceding steps. When it is not possible to apply control measures that will eliminate or reduce a hazard in a subsequent process step, then control measures must be undertaken to prevent contamination and/or growth at every individual step of the process.

Severity of a hazard							
	Death, Disability, Hospitilazation	Very serious	VS	Very low	Low	Medium	High	High
	Illness with absenteeism, Injuries, Medical treatment	Serious	S	Very low	Very low	Low	Medium	High
	Discomfort, unrest	Less serious	LS	Very low	Very low	Very low	Low	Medium
	No effect on consumers health	No effect	NE	Very low	Very low	Very low	Very low	Low
	Severity			NR	PI	S	M	G
				Not relevant	Practical impossibility	Small	Medium	Big
			Probability	No consumers affected	Less than 1 consumer affected in 100 years	2 or less consumers affected in 50 years	2 to 50 consumers affected in 50 years	More than 1 consumer affected in 1 year

Probability of a hazard

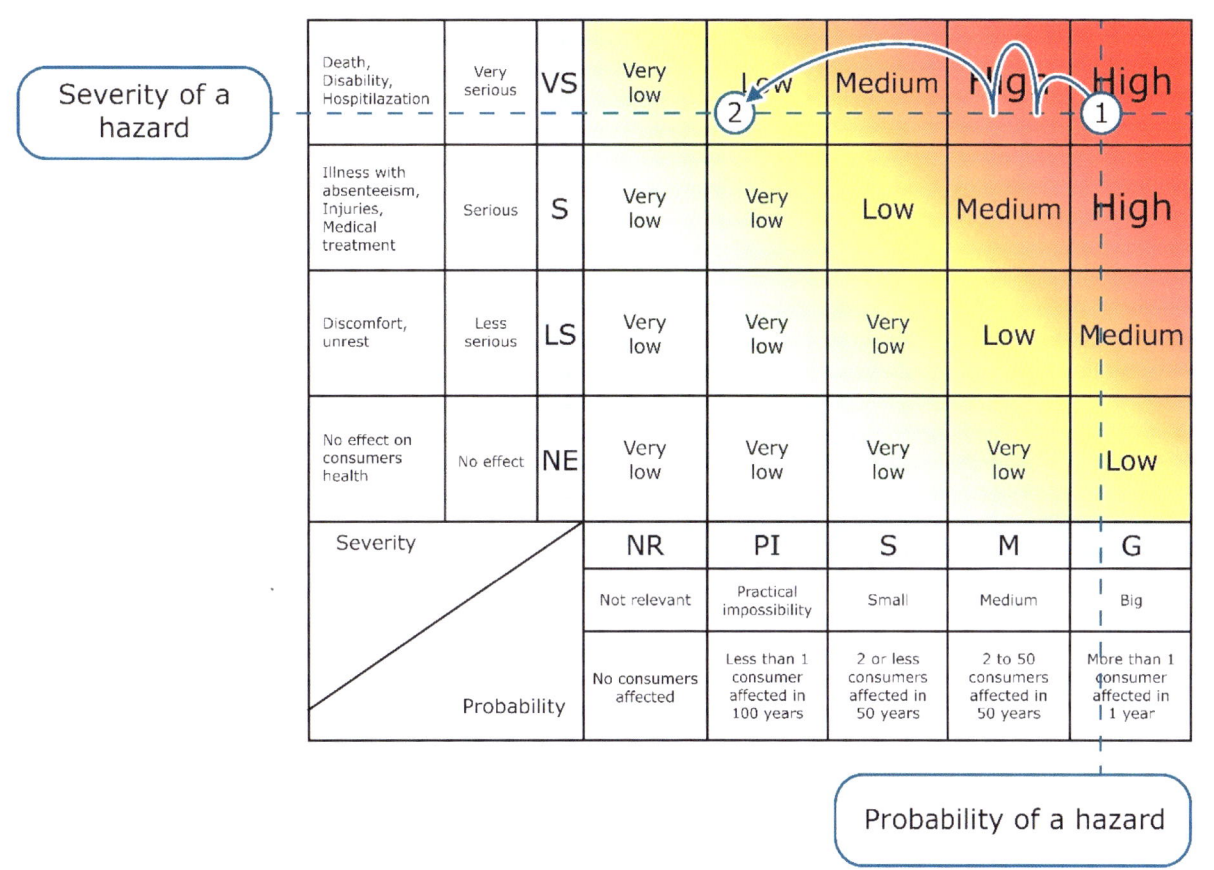

Figure 5.6 Risk matrix with control measures.

Figure 5.6 is also a good illustration of the purpose of validation. Validation of control measures, including critical limits, should demonstrates that the control measure is able to make the jump from a high to a low risk in the matrix (from point 1 to 2).

Example
"Botulinum cook"
A botulinum cook is a heat treatment applied to sterilized long shelf-life preserves (with a pH higher than 4.5) in tin and glass. The concept of the botulinum cook was coined in 1928 by the American Meyer and is widely accepted. The concept implies that the number of Clostridium bacteria cells is reduced to 12 decimal reductions or 12 D-concept. This can be achieved with a heat treatment at a core temperature of 121.1°C for the duration of 2.45 minutes. Theoretically, 12 D-concept means that a spore of Clostridium botulinum will be present in a very limited number of packages.

Reckoning example: The initial germ count in a can of one litre is set to 10,000 Clostridium botulinum (10/ml). After heat treatment the count decreased to 10^{-8}, which theoretically means that there is still contamination in one among the 100,000,000 cans. The spot on question is why a 12D-reduction is the standard and why not choose for a 16D- of a 20D-reduction that principally offers a safer product (a bigger jump to the left of the *figure). The past 70-80 years of practice has shown that there have been no reports of botulism in the products sterilized using the 12D-concept. Residual risk of 12D-concept is thus considered as low or acceptable.*

For many sterilized meat products, the botulinum cook is not applicable. The heat diffusion in solid products such as canned ham is so slow that very long holding times are needed to bring the core temperature to 121°C. The quality (flavour, aroma and colour) of the product suffer a great deal from such prolonged treatment. So a much shorter holding time is used that restricts the number of decimal reductions to for example, 4 or 5. This means that a relative high number of these type of products will still contain Clostridium botulinum. Nitrite (E 250) is added to counteract the growth of these remaining Clostridium botulinum. European legislation holds a prohibition for the use of nitrite; for reasons of food safety nitrite is permitted in a number of meat products with a long shelf life.
End of example

Once a control measure is implemented, the next question arises: what is the risk of failure of the control measure? The risk of failure of a control measure is different from the risk of the complete absence of a control measure. To illustrate this: the habit of drinking raw milk poses a risk to public health that is different from the risk of drinking pasteurized milk with the possibility of failing temperature control. In the risk matrix the failure of a control measure can be indicated by a jump to the right, from point 2 to 3 in figure 5.7. The length of this jump is determined by the frequency, the nature and the scale of deviation. A sieve that breaks every other day poses a risk that is different from the risk of a sieve that has been operating without any problems in the past five years. It is important to understand that the term risk relates to the total of the potential damage in a certain period.

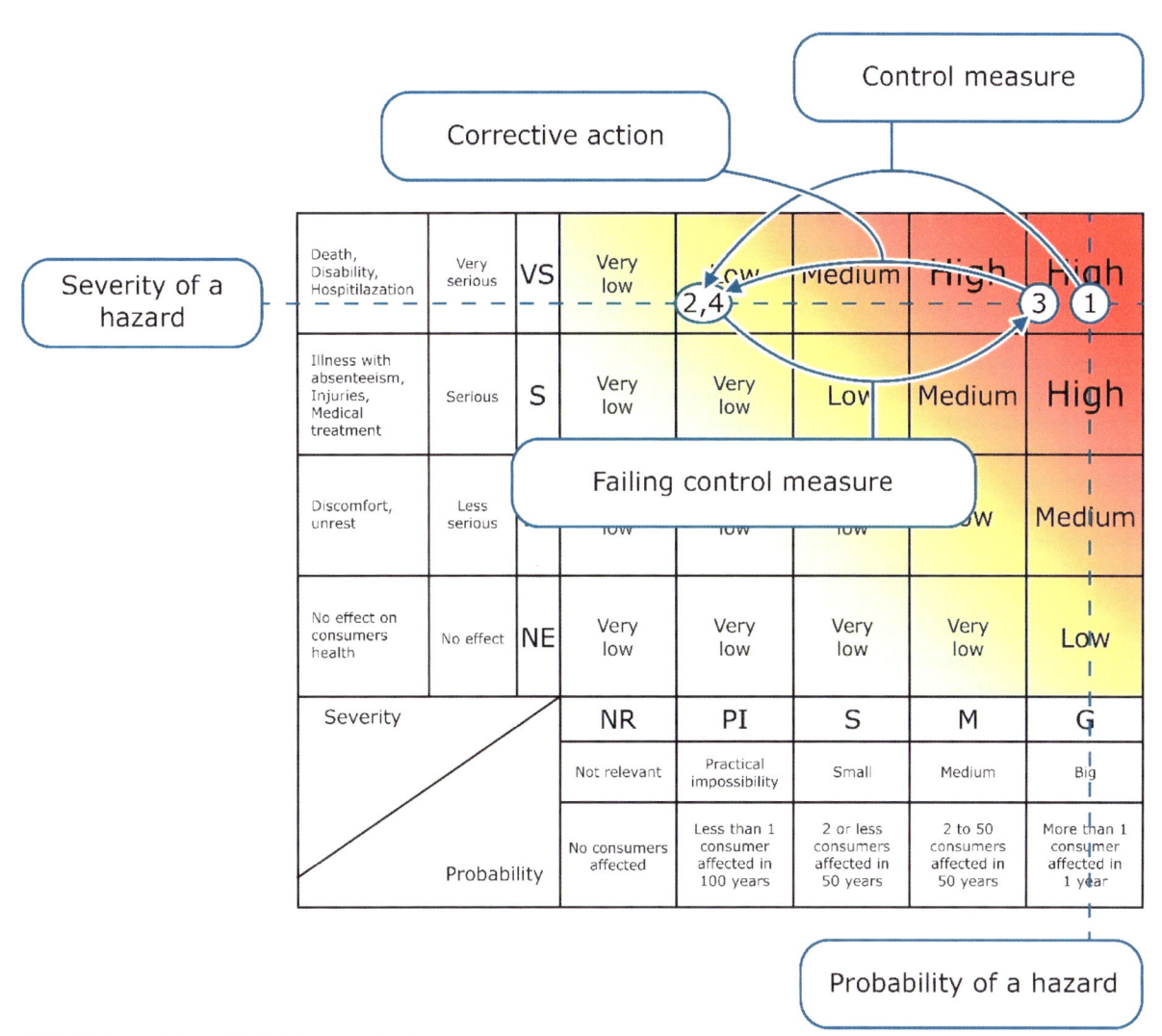

Figure 5.7 Risk matrix with failing control measure.

Monitoring procedures at CCPs should ensure that failure of control measures (the jump from 2 to 3) is detected and corrective actions should ensure that the potentially unsafe products are excluded from the process. In the risk matrix this exclusion of unsafe products is indicated by a jump to the left, from 3 to 4. When the monitoring procedure and the corrective action are well designed and properly applied, this means that point 4 coincides with point 2.

Basically, each control measure results in a jump to the left in the risk matrix, each deviation or failure leads to a jump to the right. The biggest jumps are at the security level 1, the level at which control measures are applied. The jumps become smaller in the higher levels. For example: the risk of failure of a metal detector is estimated by the probability that the failure of the detector coincides with the presence of a metal piece in a product. A risk of failure in the monitoring procedure for a metal detector is estimated by the probability that the failure of the monitoring procedure coincides with the failure of the metal detector that coincides with the presence of metal pieces in the product. A quantitative assessment of risks of deviation at higher security levels is complex, if not impossible. Nevertheless, the assessment of deviations in monitoring procedures is relevant, also see example, 'Loosening of contact during pasteurisation of milk' in section 4.8.2. A risk matrix hereby provides a useful tool.

5.2 The use of a decision tree

Experience has shown that the development of a well thought and substantiated hazard and risk analysis is a complicated process. A reason for that is food production involves a wide variety of raw materials, operations and processes. Moreover, food safety hazards are divergent in nature and health effects. It is therefore, not easy to design a decision tree which can handle this complexity. In ISO 22000 the identification of CCPs and ORPRs is based on seven aspects that are, basically an alternative to a decision tree. In the sequel of this section, these seven aspects are discussed in relation to the decision tree in HACCP guidelines.

5.2.1 CCPs and OPRPs in ISO 22000

Regarding risk assessment, ISO 22000 states in section 7.4.3 Assessment of hazards, that every food safety hazard must be assessed in accordance to the potential severity of the health consequences and the probability of their occurrence. In the following section, 7.4.4 Selection and assessment of control measures, ISO 22000 states that based on the hazard assessment in section 7.4.3, a suitable combination of control measures should be selected, that prevents, eliminates or reduces the food safety hazards to an acceptable level. In fact, this coincides with question Q1a (Is control necessary for this process step?) from the decision tree of the HACCP guidelines. The selected control measures should be then classified as measures that should be identified as OPRPs or as CCPs. The selection and classification should be done with a logical approach that includes an assessment of seven different aspects. These seven aspects that are discussed below are related to the risk of failure of the control measures. The different aspects sometimes seem to overlap each other and undoubtedly there are other interpretations possible than the interpretations outlined here. See also the table 5.4.

1. *Place within the system relative to other control measures.*

The risk of a deviation in a control measure is reduced when a there is a subsequent control measure that eliminates or reduces the hazard. In such a case the subsequent control measure takes care of the failure of the first control measure and thus the risk of failure of the first control measure is low. Consequently the first control is not essential. This line of thought can be recognized in question Q4 (Will a next step eliminate or reduce the hazard to an acceptable level?) of the decision tree in the HACCP guidelines.

2. *Whether the control measure is specifically established and applied to eliminate a or significantly reduce the level of hazard(s).*

A control that is specifically established to eliminate or reduce hazards (for example, pasteurization) basically ensures that the unsafe products are made safe. A deviation in this type of control measure (for example, a fall in temperature during pasteurization) means that certain quantity of product remains unsafe and a deviation thus rapidly leads to a high risk. In question Q2 (Is the step specifically designed to eliminate or reduce the hazard to an acceptable level?) from the decision tree in HACCP guidelines this type of control measures is considered to be CCPs. The question, whether such control measures are 'specifically established and applied' is sometimes difficult to assess. For example, is pasteurization of cheese milk intended to kill pathogenic bacteria or is it intended to create good conditions for curd production? Is a sieve specifically established and applied to remove physical contaminants or just to remove lumps of product? The distinction between control measures for safety and control measures for product quality can often be easily answered by an assessment of corresponding critical limits. The question is then whether these critical limits are based on quality criteria or food safety criteria. This is also addressed in the fourth aspect.

Unlike control measures that eliminate or reduce hazards, control measures, that prevent hazards generally have a lower risk of deviations. Control measures that prevent hazard are basically keeping safe products safe. A deviation is this type of control measure does not necessarily or directly lead to an unsafe product. The risk of deviations in control measures such as cooling, cleaning and disinfection, depends particularly on the number of deviations, the nature and its scale and the circumstances under which the deviation occurred; also see aspects 6 and 7.

3. *Synergistic effects (i.e. interaction that occurs between two or more measures resulting in their combined effect being higher than the sum of their individual effects).*

Control measures, which are aimed at excluding or limiting contamination or development of hazards, usually do not stand alone. Usually there is a combination of several different control measures. Thus, cooling of products is successful only if the contamination level of the product is below limits. Conversely, limiting contamination is effective only if growth of the pathogenic microorganisms is subsequently prevented. The risk of deviation in one of the individual control measures is restricted when the other measures function well. For example, the risk of contamination due to inadequate personal hygiene is restricted when the product is properly cooled and thus bacteria cannot grow.

4. *Its effect on identified food safety hazards relative to the strictness applied.*

There are several examples of heat treatments in which the control limits of the process are related to the quality of the product. These control limits are often at a higher temperature than the critical limits for the food safety. An example is the pasteurization of custard. The thickening of starch requires a temperature that is higher and a holding time that is longer than what is necessary for inactivation of the relevant pathogens. This is the same in case of baking bread as a deviation in time or temperature leaves the bread uncooked, which in turn leads to quality problems. In fact, these control measures must be interpreted as critical control points for the quality (QCP) and not for food safety. However, focal point here shall be that the corrective measures on QCP ensure that the products with 'somewhat lower quality' are not placed on the market as they may also be unsafe. The choice to identify a control measure as critical to quality instead of critical to safety should be guided by some thought about the distance between the critical limits for quality and the critical limits for food safety of the product.

Example:
"Deep freezing is critical for the quality"
An example of critical to quality is the storage of products in a freezer at -18°C. A failure in temperature, for example, -8°C, does not mean that the product is unsafe. In fact, the safety limits for frozen products are similar to that of refrigerated products and a critical limit for food safety applies of 7°C in combination, on condition of a limited shelf life. A failure in storage temperature for many frozen products leads to quality problems such as loss of the crystal structure of edible ice. The limit of -18°C is critical to the quality of such products, and the freezing can be thus considered as critical to quality or a QCP.

For meat products, however, freezing is directly related to food safety, especially for those meat products which are eaten raw (or pink). Freezing is a control measure to eliminate parasites such as Taenia saginata in beef and Trichinella spiralis in pork. For Trichinella in pork, maximum 15 cm thick, a critical limit applies of -15°C for 20 days or -23°C for 10 days (source: Regulation (EC) 2075/2005).
End of example

5. *Its feasibility for monitoring (e.g. ability to be monitored in a timely manner to enable immediate corrections).*

Control measures that reduce contamination are generally difficult to monitor. It is hard, if not impossible, to define measurable critical limits that could guarantee the proper application of personal hygiene. In a cleaning or disinfection process, different parameters can be measured (temperature, contact time, concentration), however, the result is mainly determined from the cleanability of the equipment. The results of disinfection as inactivation of residual bacteria can hardly be measured. Microbiological swabbing, or alternatively ATP measurement, is a possibility with limitations.

In general, control measures that prohibit or restrict the growth of bacteria can have measurable critical limits, but often have limitations. The pH of a product can be properly measured only if the product is well mixed; a deviation pertaining to inadequate mixing of an added acid can be more difficult to measure. An evident example of local variations in pH is that the growth of molds (fungi) on the surface of soft cheese ensures a local increase in the pH. A possibly present *Listeria monocytogenes* exploits this condition. For products in gas packaging, the residual oxygen content is measured and also it is tested if there is a leak in the package: the leak test, however, gives no guarantee for every individual packaging. And thus this is also a case of 'local variation'. Cooling temperatures can be monitored in storage rooms and in cargo space of a truck; it cannot be monitored in the shopping bag of a consumer.

In general, the critical limits of control measures that eliminate or reduce hazards are easier to establish and to monitor than critical limits for control measures that prevent contamination of development of hazards. The reason is that elimination or reduction of hazards often relies on a single specific process step (pasteurization, metal detection). Control measures that prohibit or restrict growth are most often applicable during several stages is the process and thus deviations can occur at different places and consequently, not easy to detect at one place. This also applies for control measures that prohibit or restrict contamination.

6. The *likelihood of failure in the functioning of a control measure or significant processing variability.*

The possibility of controlling deviations by means of monitoring and corrective actions has been discussed before. When deviations in control measures cannot always or not properly be detected and thus cannot be corrected, the risk of deviations should be reduced by lowering the probability of the causes of deviation.

Some control measures have a smaller probability of deviation than others and therefore, are more reliable. A control measure with high probability of failure, poses a greater risk to food safety than a control measure that seldom or never exceeds critical limits. A permanent magnet used for reduction of iron particles gradually loses its magnetic properties; an electromagnet directly loses its magnetic properties when the power supply is disconnected. Monitoring of an electromagnet implies verifying if the magnet is under continuous power; the testing of properties of a permanent magnet perhaps happens once or twice a year. Another example is that some sieves frequently break while other sieves function without any defect for thirty years or longer.

Example
"No monitoring on the pasteurization time"
An example of a very low probability of failure is the pasteurization time utilized by a pasteurizer for the production of drinking milk. Raw milk, to a certain extent, is contaminated with pathogenic bacteria. In the Netherlands, therefore around 70 years ago a legal obligation was enforced to pasteurize milk including a prohibition to sell raw milk in the retail channel. In a subsequent decision, the obligation for pasteurization was expanded with a requirement that the equipment used for heat treatment of raw milk must be equipped with automatic temperature control including registration and an automatic valve to dispose product that is insufficiently heated.

This example show that a control measure include monitoring and corrective action was a legal requirement long before the introduction of HACCP. Looking back one might say that the pasteurization temperature was identified as a CCP even before the concept of CCP existed.

However, for inactivation of pathogenic bacteria, along with pasteurization temperature, holding time is also important. In the regulations, however, no monitoring procedures were incorporated for pasteurization time. The underlying explanation is that the holding time is determined by the design of the pasteurization equipment. The apparatus, often a plate heat exchanger, is constructed in such a way that the minimum processing time is guaranteed. Faster flow or shorter processing times can be excluded. The probability of deviation in the pasteurization time is nil...unless someone very carelessly would increase the capacity of the equipment. Consequently, the importance of training, knowledge and understanding is once again highlighted. A very low probability of deviation means that the pasteurization time in general is not considered as a critical control point. (Partly based on DAIRY REGULATION passed in 1993).
End of example

Reducing the probability of failure of control measures is more important as the feasibility for monitoring and corrective actions decreases. Training employees is an important tool to reduce probability and magnitude of deviations. Hence, puffer fish can be slaughtered only by extremely well trained cooks, see the example in section 4.5. Training is also important, especially to prevent failure in application of personal hygiene, see the following example.

Example
"Massive food infection in North-Limburg"
On 26 and 27 April 2004, local food safety authorities were informed about an alleged food infection in a conference center in Venlo. In the period from 17 April to 25 April, out of the three groups which stayed there in the conference center, dozens of people were sick. The major symptoms were fever and watery green stools. A patient study was conducted to trace the cause. Group 1 comprised of 112 participants who participated in an international conference, 65 were sick from this group. Group 2 was a group of 23 teachers of which 3 were sick. Group 3 was a family of 20 persons, where 11 were sick. 8 out of 21 employees of the center were sick. In all 20 persons were admitted in a hospital for treatment.

Stool examination of several patients revealed the presence of Salmonella enteritidis phage type 8 with an identical DNA-pattern. Inspection of kitchen showed that kitchen staff members were also sick and one of them was already sick for a some time. This employee had been working while he was suffering from diarrhea. He specified to have vomited in the toilet near the kitchen. This toilet was also used by the guests. An inspection of the central kitchen revealed that the hygiene was not well controlled. Samples of products were taken for further investigation. In an egg salad, beet salad and rice salad, S. enteritidis phage type 8 was found. Based on the research it was concluded this explosion was most probably caused due to salads and the most suspected source of infection was the employee who had been working in spite of having diarrhea and vomiting.
(Source: Infection Disease Bulletin Volume 15, Nr. 8, 2004 and Food Safety Newsletter.)
End of example

7. The severity of the consequences of such a failure.

Next to the probability of deviation, the risk of deviations is also influenced by the nature, the scale of deviation and the circumstances under which the deviation occurs. The following example in an illustration.

Example
"Breaking the cold chain"
Breaking the cold chain is a deviation that is well measurable, but the measurement encounters several difficulties. Measurement in a refrigerator or refrigerated truck is well possible, however, becomes problematic in the shopping bag of a consumer. A monitoring system consisting of labels that change color due to a high temperature could offer a solution but generally, its necessity is not perceived to this end. The underlying idea is that a small deviation in the cooling temperature is well acceptable if it not 'too much'. Thus, the number of times the cold chain is broken should be restrained as much as possible, and it is essential to minimize the duration of the interruptions (scale of failure), especially at a high ambient temperature (circumstances under which failure occurs). In practice this means that trucks are unloaded preferably in closed dock shelters directly connected to a refrigerated unloading area. This also means that employees in companies and supermarkets along with consumers must be instructed to put the refrigerated products back in the refrigeration as soon as possible.
End of example

This concludes the discussion on the seven aspects that should be assessed in ISO 22000 while classifying control measures as OPRPs or CCPs. The similarity in the seven aspects is that they relate to the risk of failure of control measures. In the first four aspects, this risk is assessed from the perspective of nature and location of the control measure. In the last three aspects, the risk is assessed from the perspective of the possibility to correct or to prevent the failure of a control measure: the fifth aspect considers the possibility of monitoring, the sixth assesses the probability of failure of a control measure, and the seventh aspect implicates the effects of this failure. A recapitulated overview of the seven aspects is given in table 5.4

Table 5.4   Seven aspects in ISO 22000 to select OPRPs or CCPs.

	Aspect to be assessed according to section 7.4.4 of ISO 22000	When to select as Operational Prerequisite programmes?	When to select as CCP?
1	Its place within a system relative to other control measures	In a subsequent step of the process, there are control measures that can control the hazard.	Later in the process, there are no measures that can control the hazard.
2	Whether a control measure specifically established and applied to eliminate or significantly reduce the level of hazard(s).	The control measure is not specifically designed and applied to eliminate or significantly reduce the level of hazard(s). The control measure is primarily meant for realisation of certain quality characteristics of the control measure is intended to prevent hazards (by limiting contamination or growth of microorganisms).	The control measure is specifically designed and implemented to eliminate a hazard or significantly reduce it to an acceptable level.
3	Synergistic effect (i.e. interation that occurs between two or more measures resulting in their combined effect being higher than the sum of their individual effects).	There is a set of measures that ensure control.	There are several measures; however, one specific measure is decisive (essential) for total control.
4	Its effect on identified food safety hazards relative to the strictness with which a measure is applied.	Less strictness has limited consequences. The control limits of the process are far away from the confines of food safety.	Less strictness has large consequences. The control limits of the process lie close to the confines of food safety.
5	Its feasability for monitoring (e.g. ability to be monitored in a timely manner to enable immediate corrections).	Monitoring is not possible	Monitoring is certainly possible
6	The likelihood of failure in the functioning of a control measure or significant processing variability.	Low probability of deviations	High probability of deviations
7	The severity of the consequence(s) in the case of failure in its functioning.	Deviations usually do not lead to consequences for a consumer. The number of consumers possibly affected is limited	Deviations often lead to serious consequences for a consumer. The number of possibly affected consumers is substantial.

5.2.2 Decision tree based on ISO 22000.

Assessing the seven aspects in ISO 22000 is, in fact, an alternative for the use of the decision tree that is in the HACCP guidelines. A redesigned decision tree, that includes all these seven aspects will probably become too complicated. In the decision tree in figure 5.8 these aspects are combined in one question: what is the risk of failure of a control measure? This question must be answered from the aspects 2 to 7. The first aspect (place within the system with respect to other control measures) is incorporated in the answer to question 1 of the decision tree.

Question 1
Are control measures necessary for this hazard?

No, because:
- 1A: A subsequent step will eliminate or reduce the hazard to an acceptable level
- 1B: The risk is LOW

Yes, 1A and 1B are not applicable.

Ensure that all neccesary control measures are undertaken:

- **1C: Process of supplier** — Determine on the basis of risk how control at a supplier should be verified
- or **1D: Own process** — Ensure that control measures are validated
- or **1E: Process of customer** — Determine on the basis of risk in which way a customer should be informed

Question 2
Should (question 2.1) and can (question 2.2) the control measure be monitored and corrected?

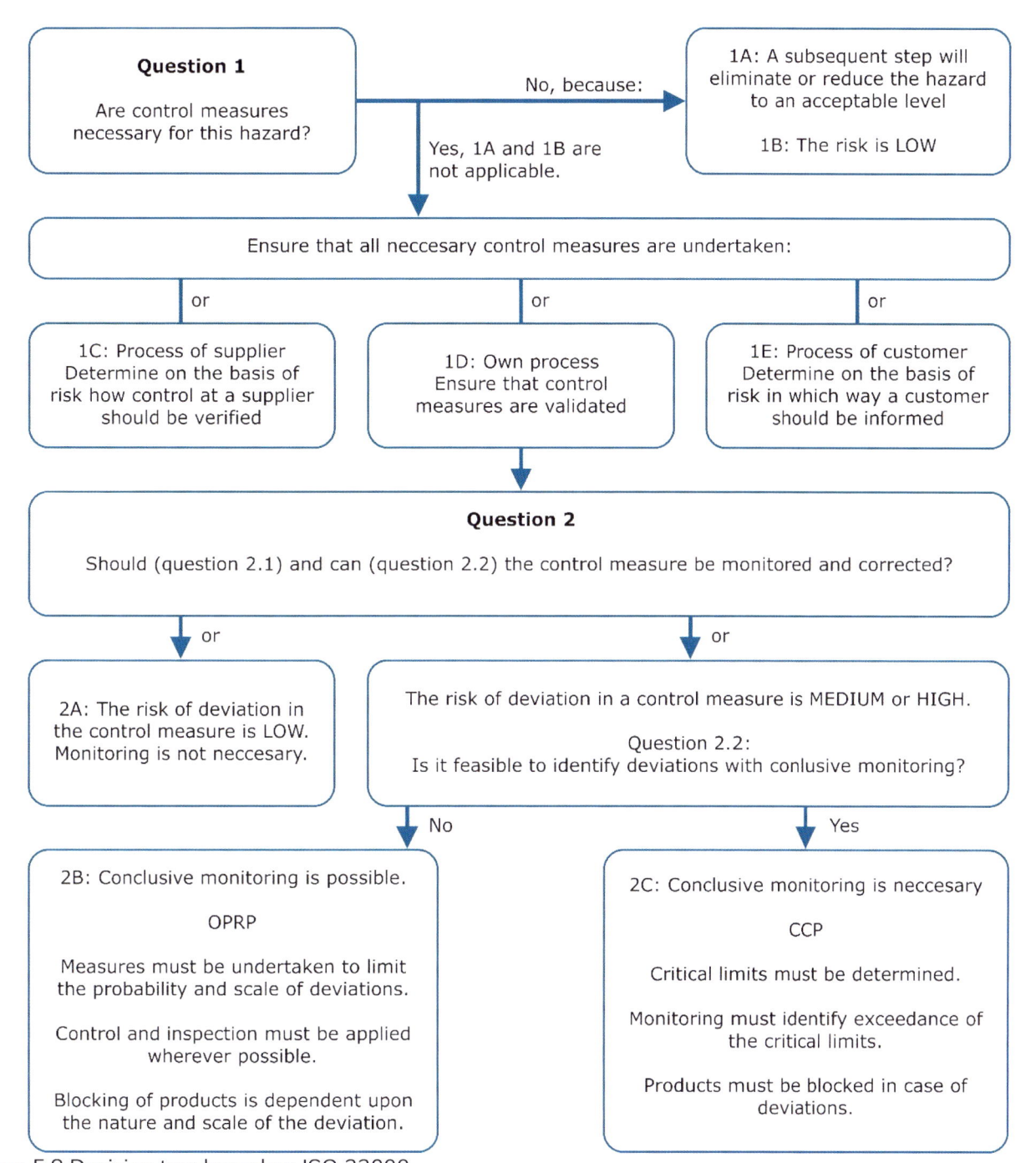

- **2A:** The risk of deviation in the control measure is LOW. Monitoring is not neccesary.
- or The risk of deviation in a control measure is MEDIUM or HIGH.
  Question 2.2: Is it feasible to identify deviations with conlusive monitoring?

No → 2B: Conclusive monitoring is possible.

OPRP

Measures must be undertaken to limit the probability and scale of deviations.

Control and inspection must be applied wherever possible.

Blocking of products is dependent upon the nature and scale of the deviation.

Yes → 2C: Conclusive monitoring is neccesary

CCP

Critical limits must be determined.

Monitoring must identify exceedance of the critical limits.

Products must be blocked in case of deviations.

Figure 5.8 Decision tree based on ISO 22000.

The decision tree in figure 5.8 focusses on two questions: whether control measures are necessary and whether monitoring of these control measures is necessary. The questions are related to the first and the second safety level in the Food Safety Bow Tie. Moreover, the decision tree is directly related to the risk matrix and also provides a basis for the control of food safety in relation to suppliers and customers. The following text explains the questions.

Question 1: Are control measures necessary for this hazard?
A risk is estimated with the use of a risk matrix where the severity of a hazard is obtained from the list of relevant hazards. The probability is determined for the situation where no control measures are implemented, also see point 1 in figure 5.6. The following answers are possible:

- 1A: There are no control measures necessary for this hazard at this step because in a subsequent process step a control measure is implemented, which will eliminate or reduce a hazard to an acceptable level.

- 1B: Risk is low thus control measures are not necessary.

When control measure are necessary there are three possibilities:
- 1C: Control measures are necessary, but it is impossible to implement them in our own process. Suppliers should undertake measures to control the hazard. The level of risk determinates how the control at the supplier should be verified, more about this is discussed in chapter 6.

- 1D: Control measures are necessary; they should be implemented in our own process. This is applicable when contamination or development can arise in our own process or when a high level of hazard is present in raw materials, and the supplier cannot ensure a sufficient control. Validation must confirm that the control measures guarantee that the probability is reduced in such a way that the risk shifts to 'low' in the risk matrix.

- 1E: Control measures are necessary, but cannot be applied in our own process or in that of the supplier. An example is *Salmonella* in raw chicken where the control typically takes place while cooking at consumers level. Another example is the presence of product-specific allergens. In these cases, the consumer must be informed about the presence of hazards. Chapter 6 more profoundly deals with this.

Question 2: Should (question 2.1) and can (question 2.2) the control measure be monitored and corrected?
Question 2.1: What is the risk of failure in a control measure?

This question deals with the reliability of a control measure: what is the probability and severity of failure? Assessment of the risk of failure in control measures is done on the basis of the risk matrix; see the jump from point 2 to point 3 in figure 5.7. In the decision tree, the risk of failure in control measures is classified as follows:

- 2A: Failure of a control measure poses a low risk and therefore, monitoring is not needed.

- Failure of a control measure poses a medium or high risk indicating a need for monitoring. This answer leads to the following question 2.2.

Question 2.2 - Can failure of the control be detected and corrected on the basis of measurable critical limits?

Here are two possible answers:
- 2B: No, monitoring and correction is not possible.

Control measures for which no measurable critical limits can be determined are identified as OPRPs. For this type of control measure, the emphasis should be on reducing the number of deviations along with their scale. The probability and the scale of deviations must be reduced in such a way that the risk of failure is acceptable. This does not mean that there is absolutely no monitoring applied in OPRPs. Regular inspections and surveillance of OPRPs can be used to identify any unfavourable developments leading to unacceptable deviations. However, due to lack of clear critical limits, this monitoring is not objective or conclusive. When an inspection of an OPRP indicates the occurrence of a deviation, then based on the nature and the scale of the deviation, it will be decided whether products should or should not be blocked.

- 2C: Yes, monitoring and correction is possible.

Monitoring procedures are introduced for control measures for which clear critical limits can be determined. Through monitoring, every exceedance of critical limits can be identified and corrective actions can be executed. An identified deviation always induces blocking of products. These are critical control points (CCPs).

5.2.3 Decision tree in the HACCP guidelines

This final section briefly discusses the relationship between the decision tree based on ISO 22000 and the decision tree based from HACCP guidelines.

Question Q1: Do control preventative measure(s) exist?
Question Q1a: Is control at this step necessary for safety?
In the decision tree in figure 5.8 primarily the order of question Q1 and question Q1a has been reversed. As compared to question Q1a, question 1 has been worded rather differently: Are control measures necessary for this hazard? The answer is 'no' (answer 1B) whenever the risk is low or (answer 1A) whenever in a subsequent process step, a measure is implemented, which will eliminate or reduce a hazard to an acceptable level. Answer 1A implies that question 4 (Will a subsequent process step eliminate or reduce a hazard to an acceptable level?) can expire.

The combination of Q1 and Q1a calls for the implementation of control measures that should but are not yet implemented. Implementation comes with modification of the step, process or product. These questions and answers are obvious - one should do what is necessary - so in the decision tree in figure 5.8 this is replaced by the simple instruction: 'Validate and implement control measures in your own process'.

Question Q2: Is the step specifically designed to eliminate or reduce a hazard to an acceptable level?
The control measures eligible for this are, for example, pasteurization, sterilization, sifting, metal detection, etc. The question here is what is the intention of Q2. In fact, question Q2 offers a shortcut whenever a question is answered as 'yes'; then question Q3 and Q4 may be skipped. This actually means that question Q2 can be excluded from the decision tree without any problems: the same results should be ultimately attained through questions Q3 and Q4.

Question Q3: Could contamination with identified hazard(s) occur in excess of acceptable level(s) or could these increase to unacceptable levels?
In this book, this question is interpreted as pertaining to the situation where a control measure fails or deviates resulting in contamination or occurrence of growth. Hence, subsequent wording is preferred for question Q3: *Can failure of a control measure lead to contamination with identified hazard(s) in excess of acceptable level(s) or could these increase to unacceptable levels?* This wording has provided a basis for question 2 *Should (question 2.1) and can (question 2.2) the control measure be monitored and corrected?*

Question Q4: Will a subsequent step eliminate or reduce a hazard to an acceptable level?
This fourth question of the decision tree from the HACCP guidelines, has been incorporated as the answer for question 1 of the decision tree in figure 5.8, see the explanation for question 1 above.

In a nutshell, this means that in the original decision tree, only two questions remain in different wordings:
Question 1 (was Q1a): Are control measures necessary for this hazard?
Question 2.1 (was Q3): What is the risk of deviation(s) in the control measure?

Chapter 6: Hazard and risk analysis in the food supply chain

In chapters 4 and 5 the principles of hazard and risk analysis and related tools were discussed. Chapter 6 deals with the practical execution. This chapter provides guidance for a hazard and risk analysis for processes, raw materials and product application. Chapter 6 also demonstrates how the hazard and risk analysis can be documented.

6.1 Control in the food supply chain

Control in the food supply chain ultimately means that there is a seamless connection between the food safety management system of suppliers, processors and clients. The HACCP principles or guidelines, however, give no specific guidance to the safety of raw materials. The HACCP guidelines state that: 'the HACCP team should list all hazards that may be reasonably expected to occur at each step from the primary production, processing, manufacture, and distribution until the point of consumption'. Since it is often impossible for individual companies to survey the entire food supply chain, this means that a hazard and risk analysis should be ideally performed in close cooperation between all the parties in the chain.

European Regulation (EC) 852/2004 (Annex II, Chapter IX) states that a food company should not accept raw materials that are known to be, or might reasonably expected to be, contaminated to an extent that even after the normal processing procedures, the final product would be unfit for human consumption. Based on this clause it can be stated that a food processor is obliged to verify which hazards could be present in raw materials from its supplier. Subsequently it could be argued that a supplier is obliged to inform clients with regard to potential hazards in the supplied products.

The ISO 22000 standard indicates that companies should ensure that sufficient information regarding all aspects of food safety is available for the entire food supply chain. To this end companies must establish, adopt and maintain effective procedures for communication with suppliers and clients concerning queries and complaints of customers and concerning product information such as intended use, storage requirements and sustainability. ISO 22000 recommends no special requirements for the hazard analysis related to suppliers and/or consumers.

Altogether, this means that the principles and definitions of HACCP form a technical approach aiming at the entire food chain. The mutual relationship between suppliers, processors and clients is a matter of division of responsibilities and this division itself is independent of the technique of hazard and risk analysis. The succeeding paragraphs describe how a company can perform a hazard and risk analysis pertaining to suppliers of raw materials and pertaining to the application of products by consumers. This approach is consistent with the decision tree based on ISO 22000 in figure 5.7. As it will be discussed in the following paragraph, risk assessment indicates how profoundly the various responsibilities of suppliers and clients should be interpreted and which measures are possible.

Figure 6.1 gives an overview of the hazard and risk analysis and demonstrates that the analysis is divided into three components: for raw materials, for processes, and for the application of products. Section 6.2 displays how the hazard and risk analysis of the processor can be written in a table that is designed on the basis of the principles of chapter 4 and the method in chapter 5. A slightly adapted method and thus a differently designed table are used for the hazard and risk analysis of the raw materials and the application of products. This is discussed in sections 6.3 and 6.4.

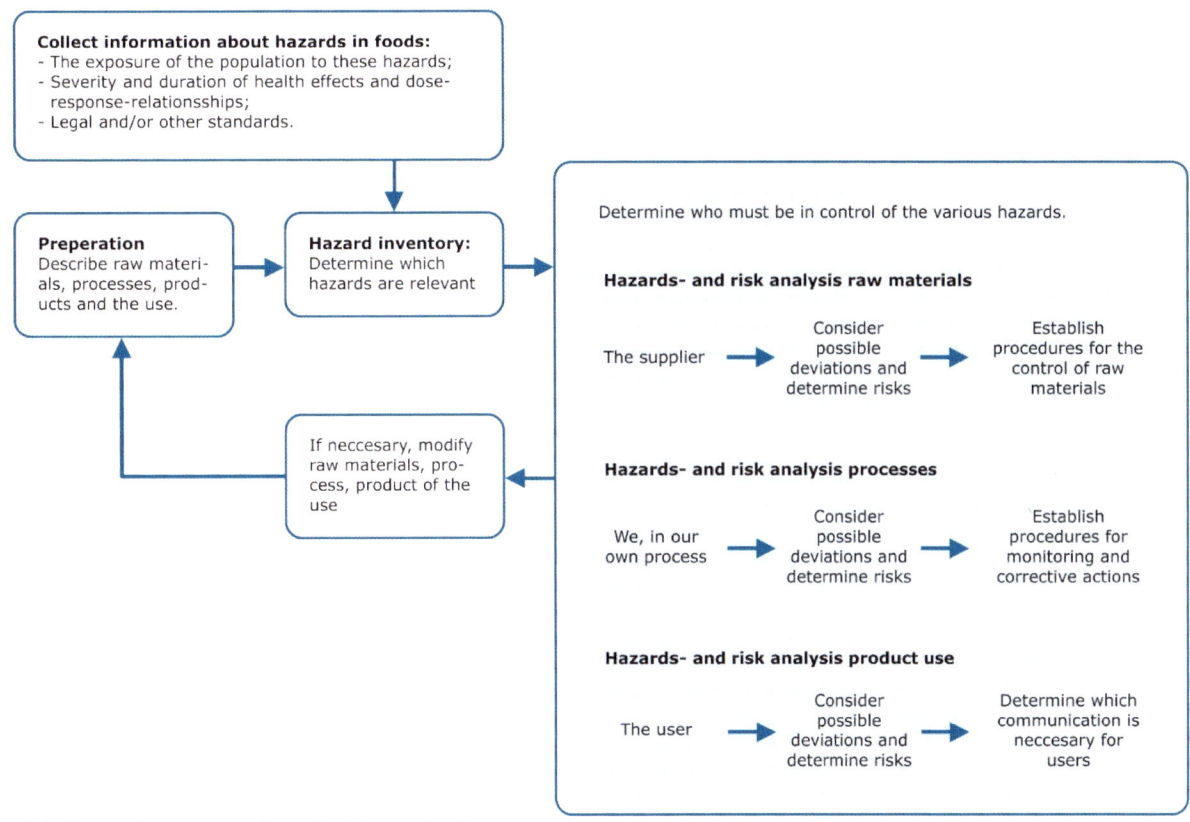

Figure 6.1 Overview hazard and risk analysis

6.2 Hazard and risk analysis of processes

A table is used to record the hazard and risk analysis. Tables 6.1 and 6.2 illustrate the completed tables. Table 6.2 is based on the example of puffer-fish cited in section 4.5. The examples demonstrate how the table can be used. For the sake of simplicity, only one hazard (glass pieces) or one process step (evisceration) has been considered.

Table 6.1 Example of a complete analysis for glass pieces in apple puree

1. Process step	2. Hazards	3. What is the risk without control measures? Severity × probability = risk			4. Explanation of risk	5. Question 1: Are control measures neccesary?	6. Question 2: Is monitoring of control measure neccesary?	7. OPRP or CCP?
Receiving	Glass	Severe	High	High	Glass coming from the chests. There emerge no glass pieces in this step.	No- 1A. Glass pieces bigger than 1 mm are removed by sieving. Ensure clean chests.		
Washing	Glass	Severe	High	High	Glass coming from the chests. There emerge no glass pieces in this step.	No- 1A. The washer removes some of the glass pieces. The remnants are removed in the process step Sieving.		
Cooking	Glass	Severe	High	High	Glass coming from the chests. There emerge no glass pieces in this step.	No- 1A. Glass pieces bigger than 1 mm are removed in the process step Sieving.		
Sieving (1mm)	Glass	Severe	High	High	Glass coming from the chests. There emerge no glass pieces in this step.	Yes- 1D. The sieve segregates all glass pieces bigger than 1mm from the product.	Yes- 2C. The sieve is made up of steel wire-mesh. The sieve rips 10 to 12 times per year.	CCP1
Bottling in jars	Glass	Severe	High	High	High probability of contamination due to the breaking of jars.	Yes- 1D. Preventing glass breakage by accurate adjustment of filling machine.	Yes- 2C. Probability of deviation (breakage) is high. Monitoring by operator (glass breakage procedure)	CCP2
Pasteuri-sation	Glass	Severe	Low	Low	Product is packed	No- 1B. Risk is low.		
Leakage detection	Idem							
Storage	Idem							

Table 6.2 Segment of an analysis for puffer-fish.

1. Process step	2. Hazard	3. What is the risk without control measures? Severity x probability = risk			4. Explanation of risk assessment	5. Question 1: Are control measures neccesary?	6. Question 2: Is monitoring of control measures neccesary?	7. OPRP or CCP?
Evisceration	Tetrodotoxin	Very severe	High	High	A small damage can lead to contamination	Yes- 1D. Contamination must be prevented by removing viscera undamaged	No- 2B. Monitoring based on critical limits is not possible. Inspections and controls must be applied; no guarantee can be obtained from this. Training to chefs is neccesary for minimising the chances of failure.	OPRP

In columns 1 and 2 of the table for each process step are recorded, which hazards could lead to contamination and which could develop. Column 2 is completed on the basis of the hazards that are included as relevant in the list of relevant hazards. Subsequently, the risk is assessed in column 3. Hereby, the severity is obtained from the list of relevant hazards. Then the probability is assessed for the situation in which no control measures were implemented. An explanation regarding risk assessment can be provided in column 4. In the subsequent columns 5 and 6, the questions are asked from the decision tree in figure 5.8. Based on the answer specified in column 6, column 7 indicates whether the control measure should be classified as OPRP or CCP.

6.3 Hazard and risk analysis of raw materials

For companies which use a limited number of raw materials and where raw materials do not pose high risks, the hazard and risk analysis of raw materials can be included as a general part in the analysis for the process. Division of hazard and risk analysis in distinct sections for raw materials and for processes becomes more effective when the number of raw materials rises and control becomes more complex. The division of the analysis then offers the opportunity to involve the responsible officers, the purchaser of raw materials and production manager of processes, directly and efficiently in the analysis.

An example of a complete analysis is shown in table 6.3. The first column of the table shows raw materials instead of process steps.

Table 6.3: Hazard and risk analysis of egg yolk

1. Process step	2. Hazard	3. What is the risk without control measures? Severity × probability = risk			4. Explanation of risk assessment	5. Question 1: Are control measures neccesary?	6. Question 2: Is monitoring of control measures neccesary?
Pasteurized egg yolk	Salmonella	Very severe	High	High	Eggs are among the most important sources of Salmonella infection. A Salmonella contamination will not be eliminated by our process.	Yes. Egg yolk must be pasteurized by the supplier.	Deviations in pasteurization can lead to the presence of Salmonella. An analysis report must be delivered with every delivered batch,

The use of columns 2, 3 and 4 is as described before. Column 4 comprises of the first question from the decision tree in figure 5.8: 'Are control measures necessary for this raw material?' Answer 1A is applicable when a hazard in a raw material will be eliminated or reduced in the process of our company, the step in which this occurs will be then a part of hazard and risk analysis of the processes of the company. Answer 1C is applicable when a hazard in raw material must be controlled in the process of the supplier. Answer 1E is applicable when a hazard in raw material can neither be controlled by the supplier nor in our own process: the control must be then left to the consumer.

In case of answer 1C (control by the supplier), the control measures are placed in column 5. The receiving company is now responsible for verifying whether the control measures are applied correctly by the supplier. The way in which this verification should be carried out is determined on the basis of the risk that the supplier has no proper control of a hazard. This is entered in column 6. This has following possibilities:

- Low risk: there should be at least a specification available for verification. Also certificates of a food safety management system can be requested.

- Medium risk: for verification, depending on situation and risk, a questionnaire related to specific hazards can be filled in. Preferably, a supplier audit is carried out and/or the results of sampling and analysis are requested. Furthermore, a company itself can ensure random sampling and analysis.

- High risk: the supplier is unable to demonstrate sufficient control. The receiving processor now must ensure control. In this case, answer 1C from the decision tree is not applicable and control measures must be undertaken in your own process. In this situation, there are two possibilities. The first is to modify the process as to implement a control measure that will eliminate or reduce the hazard. The second possibility is to implement a positive release procedure as a control measure. A positive release procedure will screen unsafe products using analysis of a representative sampling. A batch of raw materials can only be released when permitted by the results of the analysis. See figure 4.6 cited in section 4.8 and the following example from European Regulation (EC) 178/2010.

Example
"*European Regulation (EC) 178/2010 regarding sampling and analysis of mycotoxins in among others peanuts.*"
In European Regulation (EC) nr.1881/2006, maximum levels have been established for certain contaminants such as mycotoxins in foodstuffs. Sampling plays a crucial part in the precision of the determination of the levels of mycotoxins, which are heterogeneously distributed in a lot. It is therefore, necessary to establish criteria for the sampling method. The requirements for sampling are listed in the Annexes of European Regulation (EC) 178/2010. The requirements specify how many samples should be taken and what should be the size of the samples. It especially depends on the size of the batch.

It should be noted that the above regulation applies to the official controls laid down in Regulation (EC) nr.882/2004.
End of example

A problem with the hazard and risk analysis of raw materials could be that a company is not always acquainted with the control measures that are implemented by the suppliers. As the risk increases, the need for this information also increases. Practically, this means that it is important for the buyers of raw materials to be actively involved in this part of hazard and risk analysis.

6.4 Hazard and risk analysis for the use of product
There are several examples where neither the processors nor the primary producers of raw materials are able to control certain hazards. For example, the most important part of the control of *Salmonella* and *Campylobacter* in chicken meat is in hands of the final user that is often also the consumer. To make certain that these users and/or consumers are aware of the hazards in the product and the corresponding measures, suppliers must ensure adequate communication.

For companies that produce high risk products, it is advisable to perform a specific analysis of the use of the product. In a hazard and risk analysis of the application of the product it should be determined which hazards need to be controlled by the user and which measures are applicable and it should also be decide what kind of information should be provided accordingly. The control measures are placed in column 4 of the table; see the examples cited in tables 6.4 and 6.5. In column 5, the risk of failure of correct application of the control measures by the customers, is assessed. On the basis of this risk, it is decided which information should be passed on to customers. For consumer products, this information can be written on the label. For business to business products, the information can be a part of the specification. A well-known example is the warning on raw-packaged chicken meat, see section 2.2. Other examples are warnings such as 'perishable after opening', 'despite careful filleting, a bone may be left unintentionally in the fish fillet' and 'honey, not suitable for children under 1 year of age'. Furthermore, see illustration 6.2.

'Au lait cru'
Cheese made from raw milk

EXTRA RIPENED

We recommend this product as being unsuitable for consumption by pregnant women, seniors, children and immune suppressed individuals.

Figure 6.2 Au lait cru

Table 6.4 Example of hazard- and risk analysis for the use of the product – infant food.

1. Process step	2. Hazard	3. What is the risk without control measures? Severity × probability = risk			4. Explanation of risk assessment	5. Question 1: Are control measures neccesary?	6. Question 2: Is monitoring of control measures neccesary?
Baby food, ready to eat. Sterilized	Contamination and growth with pathogenic micro-organisms (e.g. Salmonella) after opening.	Very severe	High	High	Sometimes, users take out a portion from a product, for example, with a spoon. The remaining product can then be contaminated. The risk is higher when pathogene can grow.	Yes. Users must work hygienically. After opening, the product must be refridgerated. The shelf life of an opened packaging must be restricted.	On label it is indicated that - a clean spoon must be used for taking out product from the packaging. - Left over must be stored in the original packaging. - Maximum durability 3 days in refridgerator.

Table 6.5 Example of hazard and risk analysis for the use of the product – honey

1. Process step	2. Hazard	3. What is the risk without control measures? Severity x probability = risk			4. Explanation of risk assessment	5. Question 1: Are control measures neccesary?	6. Question 2: Is monitoring of control measures neccesary?
Honey	Presence of spores of Clostridium botulinum Severe	Very severe	High	High	Due to inadequately developed intestinal flora, the spores of Cl. botulinum can grow in the intestines of young children (younger than 1 year).	Control measures are not possible. Exclusion, elimination or reduction of Cl. botulinum contamination is practically not feasible.	On label it is indicated that honey is not suitable for children under 1 year of age.

Annex 1 - RECOMMENDED INTERNATIONAL CODE OF PRACTICE GENERAL PRINCIPLES OF FOOD HYGIENE CAC/RCP 1-1969, Rev. 4-2003 (1)

[note]
(1) The current version of the Recommended International Code of Practice-General Principles of Food Hygiene including Annex on Hazard Analysis and Critical Control Point (HACCP) System and Guidelines for its Application was adopted by the Codex Alimentarius Commission in 1997. Amendments regarding rinsing adopted in 1999. HACCP Guidelines were revised in 2003. The Code has been sent to all Member Nations and Associate Members of FAO and WHO as an advisory text, and it is for individual governments to decide what use they wish to make of the Guidelines.

TABLE OF CONTENTS
INTRODUCTION
SECTION I - OBJECTIVES
THE CODEX GENERAL PRINCIPLES OF FOOD HYGIENE
SECTION II - SCOPE, USE AND DEFINITION
2.1 SCOPE
2.2 USE
2.3 DEFINITIONS
SECTION III - PRIMARY PRODUCTION
3.1 ENVIRONMENTAL HYGIENE
3.2 HYGIENIC PRODUCTION OF FOOD SOURCES
3.3 HANDLING, STORAGE AND TRANSPORT
3.4 CLEANING, MAINTENANCE AND PERSONNEL HYGIENE AT PRIMARY PRODUCTION
SECTION IV - ESTABLISHMENT: DESIGN AND FACILITIES
4.1 LOCATION
4.2 PREMISES AND ROOMS
4.3 EQUIPMENT
4.4 FACILITIES
SECTION V - CONTROL OF OPERATION
5.1 CONTROL OF FOOD HAZARDS
5.2 KEY ASPECTS OF HYGIENE CONTROL SYSTEMS
5.3 INCOMING MATERIAL REQUIREMENTS
5.4 PACKAGING
5.5 WATER
5.6 MANAGEMENT AND SUPERVISION
5.7 DOCUMENTATION AND RECORDS
5.8 RECALL PROCEDURES

SECTION VI - ESTABLISHMENT: MAINTENANCE AND SANITATION
6.1 MAINTENANCE AND CLEANING
6.2 CLEANING PROGRAMMES
6.3 PEST CONTROL SYSTEMS
6.4 WASTE MANAGEMENT
6.5 MONITORING EFFECTIVENESS
SECTION VII - ESTABLISHMENT: PERSONAL HYGIENE
7.1 HEALTH STATUS
7.2 ILLNESS AND INJURIES
7.3 PERSONAL CLEANLINESS
7.4 PERSONAL BEHAVIOUR
7.5 VISITORS

SECTION VIII – TRANSPORTATION
8.1 GENERAL
8.2 REQUIREMENTS
8.3 USE AND MAINTENANCE
SECTION IX - PRODUCT INFORMATION AND CONSUMER AWARENESS
9.1 LOT IDENTIFICATION
9.2 PRODUCT INFORMATION
9.3 LABELLING
9.4 CONSUMER EDUCATION
SECTION X – TRAINING
10.1 AWARENESS AND RESPONSIBILITIES
10.2 TRAINING PROGRAMMES
10.3 INSTRUCTION AND SUPERVISION
10.4 REFRESHER TRAINING
HAZARD ANALYSIS AND CRITICAL CONTROL POINT (HACCP) SYSTEM AND GUIDELINES FOR ITS APPLICATION
PREAMBLE
DEFINITIONS
PRINCIPLES OF THE HACCP SYSTEM
GUIDELINES FOR THE APPLICATION OF THE HACCP SYSTEM
INTRODUCTION
APPLICATION
TRAINING

INTRODUCTION

People have the right to expect the food they eat to be safe and suitable for consumption. Foodborne illness and foodborne injury are at best unpleasant; at worst, they can be fatal. But there are also other consequences. Outbreaks of foodborne illness can damage trade and tourism, and lead to loss of earnings, unemployment and litigation. Food spoilage is wasteful, costly and can adversely affect trade and consumer confidence. International food trade, and foreign travel, are increasing, bringing important social and economic benefits. But this also makes the spread of illness around the world easier. Eating habits too, have undergone major change in many countries over the last two decades and new food production, preparation and distribution techniques have developed to reflect this. Effective hygiene control, therefore, is vital to avoid the adverse human health and economic consequences of foodborne illness, foodborne injury, and food spoilage. Everyone, including farmers and growers, manufacturers and processors, food handlers and consumers, has a responsibility to assure that food is safe and suitable for consumption.

These General Principles lay a firm foundation for ensuring food hygiene and should be used in conjunction with each specific code of hygienic practice, where appropriate, and the guidelines on microbiological criteria. The document follows the food chain from primary production through to final consumption, highlighting the key hygiene controls at each stage. It recommends a HACCP-based approach wherever possible to enhance food safety as described in *Hazard Analysis and Critical Control Point (HACCP) System and Guidelines for its Application* (Annex).

The controls described in this General Principles document are internationally recognized as essential to ensure the safety and suitability of food for consumption. The General Principles are commended to Governments, industry (including individual primary producers, manufacturers, processors, food service operators and retailers) and consumers alike.

SECTION I - OBJECTIVES

1.1 THE CODEX GENERAL PRINCIPLES OF FOOD HYGIENE:

identify the *essential* principles of food hygiene applicable *throughout the food chain* (including primary production through to the final consumer), to achieve the goal of ensuring that food is safe and suitable for human consumption;

- recommend a HACCP-based approach as a means to enhance food safety;
- indicate how to implement those principles; and
- provide a guidance for specific codes which may be needed for - sectors of the food chain; processes; or commodities; to amplify the hygiene requirements specific to those areas.

SECTION II - SCOPE, USE AND DEFINITION

2.1 SCOPE

2.1.1 The food chain

This document follows the food chain from primary production to the final consumer, setting out the necessary hygiene conditions for producing food which is safe and suitable for consumption. The document provides a base-line structure for other, more specific, codes applicable to particular sectors. Such specific codes and guidelines should be read in conjunction with this document and *Hazard Analysis and Critical Control Point (HACCP) System and Guidelines for its Application* (Annex).

2.1.2 Roles of Governments, industry, and consumers

Governments can consider the contents of this document and decide how best they should encourage the implementation of these general principles to:

- protect consumers adequately from illness or injury caused by food; policies need to consider the vulnerability of the population, or of different groups within the population;
- provide assurance that food is suitable for human consumption;
- maintain confidence in internationally traded food; and
- provide health education programmes which effectively communicate the principles of food hygiene to industry and consumers.

Industry should apply the hygienic practices set out in this document to:
- provide food which is safe and suitable for consumption;

- ensure that consumers have clear and easily-understood information, by way of labelling and other appropriate means, to enable them to protect their food from contamination and growth/survival of foodborne pathogens by storing, handling and preparing it correctly; and
- maintain confidence in internationally traded food.

Consumers should recognize their role by following relevant instructions and applying appropriate food hygiene measures.

2.2 USE

Each section in this document states both the objectives to be achieved and the rationale behind those objectives in terms of the safety and suitability of food.

Section III covers primary production and associated procedures. Although hygiene practices may differ considerably for the various food commodities and specific codes should be applied where appropriate, some general guidance is given in this section. Sections IV to X set down the general hygiene principles which apply throughout the food chain to the point of sale. Section IX also covers consumer information, recognizing the important role played by consumers in maintaining the safety and suitability of food.

There will inevitably be situations where some of the specific requirements contained in this document are not applicable. The fundamental question in every case is "what is necessary and appropriate on the grounds of the safety and suitability of food for consumption?"

The text indicates where such questions are likely to arise by using the phrases "where necessary" and "where appropriate". In practice, this means that, although the requirement is generally appropriate and reasonable, there will nevertheless be some situations where it is neither necessary nor appropriate on the grounds of food safety and suitability. In deciding whether a requirement is necessary or appropriate, an assessment of the risk should be made, preferably within the framework of the HACCP approach. This approach allows the requirements in this document to be flexibly and sensibly applied with a proper regard for the overall objectives of producing food which is safe and suitable for consumption. In so doing it takes into account the wide diversity of activities and varying degrees of risk involved in producing food. Additional guidance is available in specific food codes.

2.3 DEFINITIONS

For the purpose of this Code, the following expressions have the meaning stated:

Cleaning - the removal of soil, food residue, dirt, grease or other objectionable matter.
Contaminant - any biological or chemical agent, foreign matter, or other substances not intentionally added to food which may compromise food safety or suitability.
Contamination - the introduction or occurrence of a contaminant in food or food environment.
Disinfection - the reduction, by means of chemical agents and/or physical methods, of the number of micro-organisms in the environment, to a level that does not compromise food safety or suitability.
Establishment - any building or area in which food is handled and the surroundings under the control of the same management.
Food hygiene - all conditions and measures necessary to ensure the safety and suitability of food at all stages of the food chain.
Hazard - a biological, chemical or physical agent in, or condition of, food with the potential to cause an adverse health effect.
HACCP - a system which identifies, evaluates, and controls hazards which are significant for food safety.
Food handler - any person who directly handles packaged or unpackaged food, food equipment and utensils, or food contact surfaces and is therefore expected to comply with food hygiene requirements
Food safety - assurance that food will not cause harm to the consumer when it is prepared and/or eaten according to its intended use.
Food suitability - assurance that food is acceptable for human consumption according to its intended use.
Primary production - those steps in the food chain up to and including, for example, harvesting, slaughter, milking, fishing.

SECTION III - PRIMARY PRODUCTION
OBJECTIVES:

Primary production should be managed in a way that ensures that food is safe and suitable for its intended use. Where necessary, this will include:
- avoiding the use of areas where the environment poses a threat to the safety of food;
- controlling contaminants, pests and diseases of animals and plants in such a way as not to pose a threat to food safety;
- adopting practices and measures to ensure food is produced under appropriately hygienic conditions.

RATIONALE:
To reduce the likelihood of introducing a hazard which may adversely affect the safety of food, or its suitability for consumption, at later stages of the food chain.

3.1 ENVIRONMENTAL HYGIENE
Potential sources of contamination from the environment should be considered. In particular, primary food production should not be carried on in areas where the presence of potentially hazardous agents would lead to an unacceptable level of such substances in food.

3.2 HYGIENIC PRODUCTION OF FOOD SOURCES
The potential effects of primary production activities on the safety and suitability of food should be considered at all times. In particular, this includes identifying any specific points in such activities where a high probability of contamination may exist and taking specific measures to minimize that probability. The HACCP-based approach may assist in the taking of such measures - see *Hazard Analysis and Critical Control (HACCP) Point System and Guidelines for its Application* (Annex).

Producers should as far as practicable implement measures to:
- control contamination from air, soil, water, feedstuffs, fertilizers (including natural fertilizers), pesticides, veterinary drugs or any other agent used in primary production;
- control plant and animal health so that it does not pose a threat to human health through food consumption, or adversely affect the suitability of the product; and
- protect food sources from faecal and other contamination.

In particular, care should be taken to manage wastes, and store hazardous agents appropriately. On-farm programmes which achieve specific food safety goals are becoming an important part of primary production and should be encouraged.

3.3 HANDLING, STORAGE AND TRANSPORT
Procedures should be in place to:
- sort food and food ingredients to segregate material which is evidently unfit for human consumption;
- dispose of any rejected material in a hygienic manner; and
- protect food and food ingredients from contamination by pests, or by chemical, physical or microbiological contaminants or other objectionable substances during handling, storage and transport.

Care should be taken to prevent, so far as reasonably practicable, deterioration and spoilage through appropriate measures which may include controlling temperature, humidity, and/or other controls.

3.4 CLEANING, MAINTENANCE AND PERSONNEL HYGIENE AT PRIMARY PRODUCTION
Appropriate facilities and procedures should be in place to ensure that:
- any necessary cleaning and maintenance is carried out effectively; and
- an appropriate degree of personal hygiene is maintained.

SECTION IV - ESTABLISHMENT: DESIGN AND FACILITIES OBJECTIVES:
Depending on the nature of the operations, and the risks associated with them, premises, equipment and facilities should be located, designed and constructed to ensure that:
- contamination is minimized;

- design and layout permit appropriate maintenance, cleaning and disinfections and minimize air borne contamination;
- surfaces and materials, in particular those in contact with food, are non-toxic in intended use and, where necessary, suitably durable, and easy to maintain and clean;
- where appropriate, suitable facilities are available for temperature, humidity and other controls; and
- there is effective protection against pest access and harbourage.

RATIONALE:
Attention to good hygienic design and construction, appropriate location, and the provision of adequate facilities, is necessary to enable hazards to be effectively controlled.

4.1 LOCATION
4.1.1 Establishments
Potential sources of contamination need to be considered when deciding where to locate food establishments, as well as the effectiveness of any reasonable measures that might be taken to protect food. Establishments should not be located anywhere where, after considering such protective measures, it is clear that there will remain a threat to food safety or suitability. In particular, establishments should normally be located away from:
- environmentally polluted areas and industrial activities which pose a serious threat of contaminating food;
- areas subject to flooding unless sufficient safeguards are provided;
- areas prone to infestations of pests;
- areas where wastes, either solid or liquid, cannot be removed effectively.

4.1.2 Equipment
Equipment should be located so that it:
- permits adequate maintenance and cleaning;
- functions in accordance with its intended use; and
- facilitates good hygiene practices, including monitoring.

4.2 PREMISES AND ROOMS
4.2.1 Design and layout
Where appropriate, the internal design and layout of food establishments should permit good food hygiene practices, including protection against cross-contamination between and during operations by foodstuffs.

4.2.2 Internal structures and fittings
Structures within food establishments should be soundly built of durable materials and be easy to maintain, clean and where appropriate, able to be disinfected. In particular the following specific conditions should be satisfied where necessary to protect the safety and suitability of food:
- the surfaces of walls, partitions and floors should be made of impervious materials with no toxic effect in intended use;
- walls and partitions should have a smooth surface up to a height appropriate to the operation;
- floors should be constructed to allow adequate drainage and cleaning;
- ceilings and overhead fixtures should be constructed and finished to minimize the build up of dirt and condensation, and the shedding of particles;
- windows should be easy to clean, be constructed to minimize the build up of dirt and where necessary, be fitted with removable and cleanable insect-proof screens. Where necessary, windows should be fixed;
- doors should have smooth, non-absorbent surfaces, and be easy to clean and, where necessary, disinfect;
- working surfaces that come into direct contact with food should be in sound condition, durable and easy to clean, maintain and disinfect. They should be made of smooth, non-absorbent materials, and inert to the food, to detergents and disinfectants under normal operating conditions.

4.2.3 Temporary/mobile premises and vending machines
Premises and structures covered here include market stalls, mobile sales and street vending vehicles, temporary premises in which food is handled such as tents and marquees.

Such premises and structures should be sited, designed and constructed to avoid, as far as reasonably practicable, contaminating food and harbouring pests.

In applying these specific conditions and requirements, any food hygiene hazards associated with such facilities should be adequately controlled to ensure the safety and suitability of food.

4.3 EQUIPMENT
4.3.1 General
Equipment and containers (other than once-only use containers and packaging) coming into contact with food, should be designed and constructed to ensure that, where necessary, they can be adequately cleaned, disinfected and maintained to avoid the contamination of food. Equipment and containers should be made of materials with no toxic effect in intended use. Where necessary, equipment should be durable and movable or capable of being disassembled to allow for maintenance, cleaning, disinfection, monitoring and, for example, to facilitate inspection for pests.

4.3.2 Food control and monitoring equipment
In addition to the general requirements in paragraph 4.3.1, equipment used to cook, heat treat, cool, store or freeze food should be designed to achieve the required food temperatures as rapidly as necessary in the interests of food safety and suitability, and maintain them effectively. Such equipment should also be designed to allow temperatures to be monitored and controlled. Where necessary, such equipment should have effective means of controlling and monitoring humidity, air-flow and any other characteristic likely to have a detrimental effect on the safety or suitability of food. These requirements are intended to ensure that:

- harmful or undesirable micro-organisms or their toxins are eliminated or reduced to safe levels or their survival and growth are effectively controlled;
- where appropriate, critical limits established in HACCP-based plans can be monitored; and
- temperatures and other conditions necessary to food safety and suitability can be rapidly achieved and maintained.

4.3.3 Containers for waste and inedible substances
Containers for waste, by-products and inedible or dangerous substances, should be specifically identifiable, suitably constructed and, where appropriate, made of impervious material. Containers used to hold dangerous substances should be identified and, where appropriate, be lockable to prevent malicious or accidental contamination of food.

4.4 FACILITIES
4.4.1 Water supply
An adequate supply of potable water with appropriate facilities for its storage, distribution and temperature control, should be available whenever necessary to ensure the safety and suitability of food.

Potable water should be as specified in the latest edition of WHO Guidelines for Drinking Water Quality, or water of a higher standard. Non-potable water (for use in, for example, fire control, steam production, refrigeration and other similar purposes where it would not contaminate food), shall have a separate system. Non-potable water systems shall be identified and shall not connect with, or allow reflux into, potable water systems.

4.4.2 Drainage and waste disposal
Adequate drainage and waste disposal systems and facilities should be provided. They should be designed and constructed so that the risk of contaminating food or the potable water supply is avoided.

4.4.3 Cleaning
Adequate facilities, suitably designated, should be provided for cleaning food, utensils and equipment. Such facilities should have an adequate supply of hot and cold potable water where appropriate.

4.4.4 Personnel hygiene facilities and toilets
Personnel hygiene facilities should be available to ensure that an appropriate degree of personal hygiene can be maintained and to avoid contaminating food. Where appropriate, facilities should include:
- adequate means of hygienically washing and drying hands, including wash basins and a supply of hot and cold (or suitably temperature controlled) water;
- lavatories of appropriate hygienic design; and
- adequate changing facilities for personnel.

Such facilities should be suitably located and designated.

4.4.5 Temperature control
Depending on the nature of the food operations undertaken, adequate facilities should be available for heating, cooling, cooking, refrigerating and freezing food, for storing refrigerated or frozen foods, monitoring food temperatures, and when necessary, controlling ambient temperatures to ensure the safety and suitability of food.

4.4.6 Air quality and ventilation
Adequate means of natural or mechanical ventilation should be provided, in particular to:
- minimize air-borne contamination of food, for example, from aerosols and condensation droplets;
- control ambient temperatures;
- control odours which might affect the suitability of food; and
- control humidity, where necessary, to ensure the safety and suitability of food.

Ventilation systems should be designed and constructed so that air does not flow from contaminated areas to clean areas and, where necessary, they can be adequately maintained and cleaned.

4.4.7 Lighting
Adequate natural or artificial lighting should be provided to enable the undertaking to operate in a hygienic manner. Where necessary, lighting should not be such that the resulting colour is misleading. The intensity should be adequate to the nature of the operation. Lighting fixtures should, where appropriate, be protected to ensure that food is not contaminated by breakages.

4.4.8 Storage
Where necessary, adequate facilities for the storage of food, ingredients and non-food chemicals (e.g. cleaning materials, lubricants, fuels) should be provided.

Where appropriate, food storage facilities should be designed and constructed to:
- permit adequate maintenance and cleaning;
- avoid pest access and harbourage;
- enable food to be effectively protected from contamination during storage; and
- where necessary, provide an environment which minimizes the deterioration of food (e.g. by temperature and humidity control).

The type of storage facilities required will depend on the nature of the food. Where necessary, separate, secure storage facilities for cleaning materials and hazardous substances should be provided.

SECTION V - CONTROL OF OPERATION
OBJECTIVE:
To produce food which is safe and suitable for human consumption by:
- formulating design requirements with respect to raw materials, composition, processing, distribution, and consumer use to be met in the manufacture and handling of specific food items; and
- designing, implementing, monitoring and reviewing effective control systems.

RATIONALE:
To reduce the risk of unsafe food by taking preventive measures to assure the safety and suitability of food at an appropriate stage in the operation by controlling food hazards.

5.1 CONTROL OF FOOD HAZARDS
Food business operators should control food hazards through the use of systems such as HACCP. They should:
- identify any steps in their operations which are critical to the safety of food;
- implement effective control procedures at those steps;
- monitor control procedures to ensure their continuing effectiveness; and
- review control procedures periodically, and whenever the operations change.

These systems should be applied throughout the food chain to control food hygiene throughout the shelf-life of the product through proper product and process design.

Control procedures may be simple, such as checking stock rotation calibrating equipment, or correctly loading refrigerated display units. In some cases a system based on expert advice, and involving documentation, may be appropriate. A model of such a food safety system is described in *Hazard Analysis and Critical Control (HACCP) System and Guidelines for its Application* (Annex).

5.2 KEY ASPECTS OF HYGIENE CONTROL SYSTEMS
5.2.1 Time and temperature control
Inadequate food temperature control is one of the most common causes of foodborne illness or food spoilage. Such controls include time and temperature of cooking, cooling, processing and storage. Systems should be in place to ensure that temperature is controlled effectively where it is critical to the safety and suitability of food. Temperature control systems should take into account:
- the nature of the food, e.g. its water activity, pH, and likely initial level and types of micro-organisms;
- the intended shelf-life of the product;
- the method of packaging and processing; and
- how the product is intended to be used, e.g. further cooking/processing or ready-to-eat.

Such systems should also specify tolerable limits for time and temperature variations.
Temperature recording devices should be checked at regular intervals and tested for accuracy.

5.2.2 Specific process steps
Other steps which contribute to food hygiene may include, for example:
- chilling
- thermal processing
- irradiation
- drying
- chemical preservation
- vacuum or modified atmospheric packaging

5.2.3 Microbiological and other specifications
Management systems described in paragraph 5.1 offer an effective way of ensuring the safety and suitability of food. Where microbiological, chemical or physical specifications are used in any food control system, such specifications should be based on sound scientific principles and state, where appropriate, monitoring procedures, analytical methods and action limits.

5.2.4 Microbiological cross-contamination
Pathogens can be transferred from one food to another, either by direct contact or by food handlers, contact surfaces or the air. Raw, unprocessed food should be effectively separated, either physically or by time, from ready-to-eat foods, with effective intermediate cleaning and where appropriate disinfection.

Access to processing areas may need to be restricted or controlled. Where risks are particularly high, access to processing areas should be only via a changing facility. Personnel may need to be required to put on clean protective clothing including footwear and wash their hands before entering.

Surfaces, utensils, equipment, fixtures and fittings should be thoroughly cleaned and where necessary disinfected after raw food, particularly meat and poultry, has been handled or processed.

5.2.5 Physical and chemical contamination
Systems should be in place to prevent contamination of foods by foreign bodies such as glass or metal shards from machinery, dust, harmful fumes and unwanted chemicals. In manufacturing and processing, suitable detection or screening devices should be used where necessary.

5.3 INCOMING MATERIAL REQUIREMENTS
No raw material or ingredient should be accepted by an establishment if it is known to contain parasites, undesirable micro-organisms, pesticides, veterinary drugs or toxic, decomposed or extraneous substances which would not be reduced to an acceptable level by normal sorting and/or processing. Where appropriate, specifications for raw materials should be identified and applied.

Raw materials or ingredients should, where appropriate, be inspected and sorted before processing. Where necessary, laboratory tests should be made to establish fitness for use. Only sound, suitable raw materials or ingredients should be used.

Stocks of raw materials and ingredients should be subject to effective stock rotation.

5.4 PACKAGING
Packaging design and materials should provide adequate protection for products to minimize contamination, prevent damage, and accommodate proper labelling. Packaging materials or gases where used must be non-toxic and not pose a threat to the safety and suitability of food under the specified conditions of storage and

use. Where appropriate, reusable packaging should be suitably durable, easy to clean and, where necessary, disinfect.

5.5 WATER
5.5.1 In contact with food
Only potable water, should be used in food handling and processing, with the following exceptions:
- for steam production, fire control and other similar purposes not connected with food; and
- in certain food processes, e.g. chilling, and in food handling areas, provided this does not constitute a hazard to the safety and suitability of food (e.g. the use of clean sea water).

Water recirculated for reuse should be treated and maintained in such a condition that no risk to the safety and suitability of food results from its use. The treatment process should be effectively monitored. Recirculated water which has received no further treatment and water recovered from processing of food by evaporation or drying may be used, provided its use does not constitute a risk to the safety and suitability of food.

5.5.2 As an ingredient
Potable water should be used wherever necessary to avoid food contamination.

5.5.3 Ice and steam
Ice should be made from water that complies with section 4.4.1. Ice and steam should be produced, handled and stored to protect them from contamination.

Steam used in direct contact with food or food contact surfaces should not constitute a threat to the safety and suitability of food.

5.6 MANAGEMENT AND SUPERVISION
The type of control and supervision needed will depend on the size of the business, the nature of its activities and the types of food involved. Managers and supervisors should have enough knowledge of food hygiene principles and practices to be able to judge potential risks, take appropriate preventive and corrective action, and ensure that effective monitoring and supervision takes place.

5.7 DOCUMENTATION AND RECORDS
Where necessary, appropriate records of processing, production and distribution should be kept and retained for a period that exceeds the shelf-life of the product. Documentation can enhance the credibility and effectiveness of the food safety control system.

5.8 RECALL PROCEDURES
Managers should ensure effective procedures are in place to deal with any food safety hazard and to enable the complete, rapid recall of any implicated lot of the finished food from the market. Where a product has been withdrawn because of an immediate health hazard, other products which are produced under similar conditions, and which may present a similar hazard to public health, should be evaluated for safety and may need to be withdrawn. The need for public warnings should be considered.

Recalled products should be held under supervision until they are destroyed, used for purposes other than human consumption, determined to be safe for human consumption, or reprocessed in a manner to ensure their safety.

SECTION VI - ESTABLISHMENT: MAINTENANCE AND SANITATION
OBJECTIVE:
To establish effective systems to:
- ensure adequate and appropriate maintenance and cleaning;
- control pests;
- manage waste; and
- monitor effectiveness of maintenance and sanitation procedures.

RATIONALE:
To facilitate the continuing effective control of food hazards, pests, and other agents likely to contaminate food.

6.1 MAINTENANCE AND CLEANING
6.1.1 General
Establishments and equipment should be kept in an appropriate state of repair and condition to:
- facilitate all sanitation procedures;
- function as intended, particularly at critical steps (see paragraph 5.1);
- prevent contamination of food, e.g. from metal shards, flaking plaster, debris and chemicals.

Cleaning should remove food residues and dirt which may be a source of contamination. The necessary cleaning methods and materials will depend on the nature of the food business. Disinfection may be necessary after cleaning. Cleaning chemicals should be handled and used carefully and in accordance with manufacturers' instructions and stored, where necessary, separated from food, in clearly identified containers to avoid the risk of contaminating food.

6.1.2 Cleaning procedures and methods
Cleaning can be carried out by the separate or the combined use of physical methods, such as heat, scrubbing, turbulent flow, vacuum cleaning or other methods that avoid the use of water, and chemical methods using detergents, alkalis or acids.

Cleaning procedures will involve, where appropriate:
- removing gross debris from surfaces;
- applying a detergent solution to loosen soil and bacterial film and hold them in solution or suspension;
- rinsing with water which complies with section 4, to remove loosened soil and residues of detergent;
- dry cleaning or other appropriate methods for removing and collecting residues and debris; and
- where necessary, disinfection with subsequent rinsing unless the manufacturers' instructions indicate on scientific basis that rinsing is not required.

6.2 CLEANING PROGRAMMES
Cleaning and disinfection programmes should ensure that all parts of the establishment are appropriately clean, and should include the cleaning of cleaning equipment.

Cleaning and disinfection programmes should be continually and effectively monitored for their suitability and effectiveness and where necessary, documented.

Where written cleaning programmes are used, they should specify:
- areas, items of equipment and utensils to be cleaned;
- responsibility for particular tasks;
- method and frequency of cleaning; and
- monitoring arrangements.

Where appropriate, programmes should be drawn up in consultation with relevant specialist expert advisors.

6.3 PEST CONTROL SYSTEMS
6.3.1 General
Pests pose a major threat to the safety and suitability of food. Pest infestations can occur where there are breeding sites and a supply of food. Good hygiene practices should be employed to avoid creating an environment conducive to pests. Good sanitation, inspection of incoming materials and good monitoring can minimize the likelihood of infestation and thereby limit the need for pesticides.

6.3.2 Preventing access
Buildings should be kept in good repair and condition to prevent pest access and to eliminate potential breeding sites. Holes, drains and other places where pests are likely to gain access should be kept sealed. Wire mesh screens, for example on open windows, doors and ventilators, will reduce the problem of pest entry. Animals should, wherever possible, be excluded from the grounds of factories and food processing plants.

6.3.3 Harbourage and infestation
The availability of food and water encourages pest harbourage and infestation. Potential food sources should be stored in pest-proof containers and/or stacked above the ground and away from walls. Areas both inside and outside food premises should be kept clean. Where appropriate, refuse should be stored in covered, pest-proof containers.

6.3.4 Monitoring and detection
Establishments and surrounding areas should be regularly examined for evidence of infestation.

6.3.5 Eradication
Pest infestations should be dealt with immediately and without adversely affecting food safety or suitability. Treatment with chemical, physical or biological agents should be carried out without posing a threat to the safety or suitability of food.

6.4 WASTE MANAGEMENT

Suitable provision must be made for the removal and storage of waste. Waste must not be allowed to accumulate in food handling, food storage, and other working areas and the adjoining environment except so far as is unavoidable for the proper functioning of the business.
Waste stores must be kept appropriately clean.

6.5 MONITORING EFFECTIVENESS

Sanitation systems should be monitored for effectiveness, periodically verified by means such as audit pre-operational inspections or, where appropriate, microbiological sampling of environment and food contact surfaces and regularly reviewed and adapted to reflect changed circumstances.

SECTION VII - ESTABLISHMENT: PERSONAL HYGIENE

OBJECTIVES:
To ensure that those who come directly or indirectly into contact with food are not likely to contaminate food by:
- maintaining an appropriate degree of personal cleanliness;
- behaving and operating in an appropriate manner.

RATIONALE:
People who do not maintain an appropriate degree of personal cleanliness, who have certain illnesses or conditions or who behave inappropriately, can contaminate food and transmit illness to consumers.

7.1 HEALTH STATUS

People known, or suspected, to be suffering from, or to be a carrier of a disease or illness likely to be transmitted through food, should not be allowed to enter any food handling area if there is a likelihood of their contaminating food. Any person so affected should immediately report illness or symptoms of illness to the management.
Medical examination of a food handler should be carried out if clinically or epidemiologically indicated.

7.2 ILLNESS AND INJURIES

Conditions which should be reported to management so that any need for medical examination and/or possible exclusion from food handling can be considered, include:
- jaundice;
- diarrhea;
- vomiting;
- fever;
- sore throat with fever;
- visibly infected skin lesions (boils, cuts, etc.);
- discharges from the ear, eye or nose.

7.3 PERSONAL CLEANLINESS

Food handlers should maintain a high degree of personal cleanliness and, where appropriate, wear suitable protective clothing, head covering, and footwear. Cuts and wounds, where personnel are permitted to continue working, should be covered by suitable waterproof dressings.
Personnel should always wash their hands when personal cleanliness may affect food safety, for example:
- at the start of food handling activities;
- immediately after using the toilet; and
- after handling raw food or any contaminated material, where this could result in contamination of other food items; they should avoid handling ready-to-eat food, where appropriate.

7.4 PERSONAL BEHAVIOUR

People engaged in food handling activities should refrain from behaviour which could result in contamination of food, for example:
- smoking;
- spitting;
- chewing or eating;
- sneezing or coughing over unprotected food.
Personal effects such as jewellery, watches, pins or other items should not be worn or brought into food handling areas if they pose a threat to the safety and suitability of food.

7.5 VISITORS

Visitors to food manufacturing, processing or handling areas should, where appropriate, wear protective clothing and adhere to the other personal hygiene provisions in this section.

SECTION VIII - TRANSPORTATION

OBJECTIVES:

Measures should be taken where necessary to:
- protect food from potential sources of contamination;
- protect food from damage likely to render the food unsuitable for consumption; and
- provide an environment which effectively controls the growth of pathogenic or spoilage micro-organisms and the production of toxins in food.

RATIONALE:

Food may become contaminated, or may not reach its destination in a suitable condition for consumption, unless effective control measures are taken during transport, even where adequate hygiene control measures have been taken earlier in the food chain.

8.1 GENERAL

Food must be adequately protected during transport. The type of conveyances or containers required depends on the nature of the food and the conditions under which it has to be transported.

8.2 REQUIREMENTS

Where necessary, conveyances and bulk containers should be designed and constructed so that they:
- do not contaminate foods or packaging;
- can be effectively cleaned and, where necessary, disinfected;
- permit effective separation of different foods or foods from non-food items where necessary during transport;
- provide effective protection from contamination, including dust and fumes;
- can effectively maintain the temperature, humidity, atmosphere and other conditions necessary to protect food from harmful or undesirable microbial growth and deterioration likely to render it unsuitable for consumption; and
- allow any necessary temperature, humidity and other conditions to be checked.

8.3 USE AND MAINTENANCE

Conveyances and containers for transporting food should be kept in an appropriate state of cleanliness, repair and condition. Where the same conveyance or container is used for transporting different foods, or non-foods, effective cleaning and, where necessary, disinfection should take place between loads. Where appropriate, particularly in bulk transport, containers and conveyances should be designated and marked for food use only and be used only for that purpose.

SECTION IX - PRODUCT INFORMATION AND CONSUMER AWARENESS

OBJECTIVES:

Products should bear appropriate information to ensure that:
- adequate and accessible information is available to the next person in the food chain to enable them to handle, store, process, prepare and display the product safely and correctly;
- the lot or batch can be easily identified and recalled if necessary.

Consumers should have enough knowledge of food hygiene to enable them to:
- understand the importance of product information;
- make informed choices appropriate to the individual; and
- prevent contamination and growth or survival of foodborne pathogens by storing, preparing and using it correctly.

Information for industry or trade users should be clearly distinguishable from consumer information, particularly on food labels.

RATIONALE:

Insufficient product information, and/or inadequate knowledge of general food hygiene, can lead to products being mishandled at later stages in the food chain. Such mishandling can result in illness, or products becoming unsuitable for consumption, even where adequate hygiene control measures have been taken earlier in the food chain.

9.1 LOT IDENTIFICATION

Lot identification is essential in product recall and also helps effective stock rotation. Each container of food should be permanently marked to identify the producer and the lot. Codex General Standard for the Labelling of Prepackaged Foods (CODEX STAN 1-1985, Rev. 1(1991)) applies.

9.2 PRODUCT INFORMATION
All food products should be accompanied by or bear adequate information to enable the next person in the food chain to handle, display, store and prepare and use the product safely and correctly.

9.3 LABELLING
Prepackaged foods should be labelled with clear instructions to enable the next person in the food chain to handle, display, store and use the product safely. Codex General Standard for the Labelling of Prepackaged Foods (CODEX STAN 1-1985, Rev. (1991)) applies.

9.4 CONSUMER EDUCATION
Health education programmes should cover general food hygiene. Such programmes should enable consumers to understand the importance of any product information and to follow any instructions accompanying products, and make informed choices. In particular consumers should be informed of the relationship between time/temperature control and foodborne illness.

SECTION X - TRAINING
OBJECTIVE:
Those engaged in food operations who come directly or indirectly into contact with food should be trained, and/or instructed in food hygiene to a level appropriate to the operations they are to perform.

RATIONALE:
Training is fundamentally important to any food hygiene system.
Inadequate hygiene training, and/or instruction and supervision of *all* people involved in food related activities pose a potential threat to the safety of food and its suitability for consumption.

10.1 AWARENESS AND RESPONSIBILITIES
Food hygiene training is fundamentally important. All personnel should be aware of their role and responsibility in protecting food from contamination or deterioration. Food handlers should have the necessary knowledge and skills to enable them to handle food hygienically. Those who handle strong cleaning chemicals or other potentially hazardous chemicals should be instructed in safe handling techniques.

10.2 TRAINING PROGRAMMES
Factors to take into account in assessing the level of training required include:
- the nature of the food, in particular its ability to sustain growth of pathogenic or spoilage micro-organisms;
- the manner in which the food is handled and packed, including the probability of contamination;
- the extent and nature of processing or further preparation before final consumption;
- the conditions under which the food will be stored; and
- the expected length of time before consumption.

10.3 INSTRUCTION AND SUPERVISION
Periodic assessments of the effectiveness of training and instruction programmes should be made, as well as routine supervision and checks to ensure that procedures are being carried out effectively.
Managers and supervisors of food processes should have the necessary knowledge of food hygiene principles and practices to be able to judge potential risks and take the necessary action to remedy deficiencies.

10.4 REFRESHER TRAINING
Training programmes should be routinely reviewed and updated where necessary. Systems should be in place to ensure that food handlers remain aware of all procedures necessary to maintain the safety and suitability of food.

Annex 2 - HAZARD ANALYSIS AND CRITICAL CONTROL POINT (HACCP) SYSTEM AND GUIDELINES FOR ITS APPLICATION

Annex to CAC/RCP 1-1969 (Rev. 4 - 2003)

PREAMBLE

The first section of this document sets out the principles of the Hazard Analysis and Critical Control Point (HACCP) system adopted by the Codex Alimentarius Commission. The second section provides general guidance for the application of the system while recognizing that the details of application may vary depending on the circumstances of the food operation (2).

[note]
(2) *The Principles of the HACCP System set the basis for the requirements for the application of HACCP, while the Guidelines for the Application provide general guidance for practical application.*

The HACCP system, which is science based and systematic, identifies specific hazards and measures for their control to ensure the safety of food. HACCP is a tool to assess hazards and establish control systems that focus on prevention rather than relying mainly on end product testing. Any HACCP system is capable of accommodating change, such as advances in equipment design, processing procedures or technological developments.

HACCP can be applied throughout the food chain from primary production to final consumption and its implementation should be guided by scientific evidence of risks to human health. As well as enhancing food safety, implementation of HACCP can provide other significant benefits. In addition, the application of HACCP systems can aid inspection by regulatory authorities and promote international trade by increasing confidence in food safety.

The successful application of HACCP requires the full commitment and involvement of management and the work force. It also requires a multidisciplinary approach; this multidisciplinary approach should include, when appropriate, expertise in agronomy, veterinary health, production, microbiology, medicine, public health, food technology, environmental health, chemistry and engineering, according to the particular study. The application of HACCP is compatible with the implementation of quality management systems, such as the ISO 9000 series, and is the system of choice in the management of food safety within such systems.

While the application of HACCP to food safety was considered here, the concept can be applied to other aspects of food quality.

DEFINITIONS

Control (verb): To take all necessary actions to ensure and maintain compliance with criteria established in the HACCP plan.
Control (noun): The state wherein correct procedures are being followed and criteria are being met.
Control measure: Any action and activity that can be used to prevent or eliminate a food safety hazard or reduce it to an acceptable level.
Corrective action: Any action to be taken when the results of monitoring at the CCP indicate a loss of control.
Critical Control Point (CCP): A step at which control can be applied and is essential to prevent or eliminate a food safety hazard or reduce it to an acceptable level.
Critical limit: A criterion which separates acceptability from unacceptability.
Deviation: Failure to meet a critical limit.
Flow diagram: A systematic representation of the sequence of steps or operations used in the production or manufacture of a particular food item.
HACCP: A system which identifies, evaluates, and controls hazards which are significant for food safety.
HACCP plan: A document prepared in accordance with the principles of HACCP to ensure control of hazards which are significant for food safety in the segment of the food chain under consideration.
Hazard: A biological, chemical or physical agent in, or condition of, food with the potential to cause an adverse health effect.
Hazard analysis: The process of collecting and evaluating information on hazards and conditions leading to their presence to decide which are significant for food safety and therefore should be addressed in the HACCP plan.
Monitor: The act of conducting a planned sequence of observations or measurements of control parameters to assess whether a CCP is under control.

Step: A point, procedure, operation or stage in the food chain including raw materials, from primary production to final consumption.
Validation: Obtaining evidence that the elements of the HACCP plan are effective.
Verification: The application of methods, procedures, tests and other evaluations, in addition to monitoring to determine compliance with the HACCP plan.

PRINCIPLES OF THE HACCP SYSTEM
The HACCP system consists of the following seven principles:
PRINCIPLE 1   Conduct a hazard analysis.
PRINCIPLE 2   Determine the Critical Control Points (CCPs).
PRINCIPLE 3   Establish critical limit(s).
PRINCIPLE 4   Establish a system to monitor control of the CCP.
PRINCIPLE 5   Establish the corrective action to be taken when monitoring indicates that a particular CCP is not under control.
PRINCIPLE 6   Establish procedures for verification to confirm that the HACCP system is working effectively.
PRINCIPLE 7   Establish documentation concerning all procedures and records appropriate to these principles and their application.

GUIDELINES FOR THE APPLICATION OF THE HACCP SYSTEM
INTRODUCTION
Prior to application of HACCP to any sector of the food chain, that sector should have in place prerequisite programs such as good hygienic practices according to the Codex General Principles of Food Hygiene, the appropriate Codex Codes of Practice, and appropriate food safety requirements. These prerequisite programs to HACCP, including training, should be well established, fully operational and verified in order to facilitate the successful application and implementation of the HACCP system.

For all types of food business, management awareness and commitment is necessary for implementation of an effective HACCP system. The effectiveness will also rely upon management and employees having the appropriate HACCP knowledge and skills.

During hazard identification, evaluation, and subsequent operations in designing and applying HACCP systems, consideration must be given to the impact of raw materials, ingredients, food manufacturing practices, role of manufacturing processes to control hazards, likely end-use of the product, categories of consumers of concern, and epidemiological evidence relative to food safety.

The intent of the HACCP system is to focus control at Critical Control Points (CCPs). Redesign of the operation should be considered if a hazard which must be controlled is identified but no CCPs are found.

HACCP should be applied to each specific operation separately. CCPs identified in any given example in any Codex Code of Hygienic Practice might not be the only ones identified for a specific application or might be of a different nature. The HACCP application should be reviewed and necessary changes made when any modification is made in the product, process, or any step.

The application of the HACCP principles should be the responsibility of each individual businesses. However, it is recognized by governments and businesses that there may be obstacles that hinder the effective application of the HACCP principles by individual business. This is particularly relevant in small and/or less developed businesses. While it is recognized that when applying HACCP, flexibility appropriate to the business is important, all seven principles must be applied in the HACCP system. This flexibility should take into account the nature and size of the operation, including the human and financial resources, infrastructure, processes, knowledge and practical constraints.

Small and/or less developed businesses do not always have the resources and the necessary expertise on site for the development and implementation of an effective HACCP plan. In such situations, expert advice should be obtained from other sources, which may include: trade and industry associations, independent experts and regulatory authorities. HACCP literature and especially sector-specific HACCP guides can be valuable. HACCP guidance developed by experts relevant to the process or type of operation may provide a useful tool for businesses in designing and implementing the HACCP plan. Where businesses are using expertly developed HACCP guidance, it is essential that it is specific to the foods and/or processes under consideration. More detailed information on the obstacles in implementing HACCP, particularly in reference to SLDBs, and recommendations in resolving these obstacles, can be found in "Obstacles to the Application of HACCP, Particularly in Small and Less Developed Businesses, and Approaches to Overcome Them" (document in preparation by FAO/WHO).

The efficacy of any HACCP system will nevertheless rely on management and employees having the appropriate HACCP knowledge and skills, therefore ongoing training is necessary for all levels of employees and managers, as appropriate.

APPLICATION

The application of HACCP principles consists of the following tasks as identified in the Logic Sequence for Application of HACCP (Diagram 1).

1. Assemble HACCP team

The food operation should assure that the appropriate product specific knowledge and expertise is available for the development of an effective HACCP plan. Optimally, this may be accomplished by assembling a multidisciplinary team. Where such expertise is not available on site, expert advice should be obtained from other sources, such as, trade and industry associations, independent experts, regulatory authorities, HACCP literature and HACCP guidance (including sector-specific HACCP guides). It may be possible that a well-trained individual with access to such guidance is able to implement HACCP in-house. The scope of the HACCP plan should be identified. The scope should describe which segment of the food chain is involved and the general classes of hazards to be addressed (e.g. does it cover all classes of hazards or only selected classes).

2. Describe product

A full description of the product should be drawn up, including relevant safety information such as: composition, physical/chemical structure (including Aw, pH, etc), microcidal/static treatments (heat-treatment, freezing, brining, smoking, etc), packaging, durability and storage conditions and method of distribution. Within businesses with multiple products, for example, catering operations, it may be effective to group products with similar characteristics or processing steps, for the purpose of development of the HACCP plan.

3. Identify intended use

The intended use should be based on the expected uses of the product by the end user or consumer. In specific cases, vulnerable groups of the population, e.g. institutional feeding, may have to be considered.

4. Construct flow diagram

The flow diagram should be constructed by the HACCP team (see also paragraph 1 above). The flow diagram should cover all steps in the operation for a specific product. The same flow diagram may be used for a number of products that are manufactured using similar processing steps. When applying HACCP to a given operation, consideration should be given to steps preceding and following the specified operation.

5. On-site confirmation of flow diagram

Steps must be taken to confirm the processing operation against the flow diagram during all stages and hours of operation and amend the flow diagram where appropriate. The confirmation of the flow diagram should be performed by a person or persons with sufficient knowledge of the processing operation.

6. List all potential hazards associated with each step, conduct a hazard analysis, and consider any measures to control identified hazards - (SEE PRINCIPLE 1)

The HACCP team (see "assemble HACCP team" above) should list all of the hazards that may be reasonably expected to occur at each step according to the scope from primary production, processing, manufacture, and distribution until the point of consumption.

The HACCP team (see "assemble HACCP team") should next conduct a hazard analysis to identify for the HACCP plan, which hazards are of such a nature that their elimination or reduction to acceptable levels is essential to the production of a safe food.

In conducting the hazard analysis, wherever possible the following should be included:
- the likely occurrence of hazards and severity of their adverse health effects;

- the qualitative and/or quantitative evaluation of the presence of hazards;
- survival or multiplication of micro-organisms of concern;
- production or persistence in foods of toxins, chemicals or physical agents; and,
- conditions leading to the above.

Consideration should be given to what control measures, if any exist, can be applied to each hazard. More than one control measure may be required to control a specific hazard(s) and more than one hazard may be controlled by a specified control measure.

7. Determine Critical Control Points - (SEE PRINCIPLE 2) (3)

There may be more than one CCP at which control is applied to address the same hazard. The determination of a CCP in the HACCP system can be facilitated by the application of a decision tree (e.g., Diagram 2), which indicates a logic reasoning approach. Application of a decision tree should be flexible, given whether the operation is for production, slaughter, processing, storage, distribution or other. It should be used for guidance when determining CCPs. This example of a decision tree may not be applicable to all situations. Other approaches may be used. Training in the application of the decision tree is recommended.

If a hazard has been identified at a step where control is necessary for safety, and no control measure exists at that step, or any other, then the product or process should be modified at that step, or at any earlier or later stage, to include a control measure.

[note]
(3) Since the publication of the decision tree by Codex, its use has been implemented many times for training purposes. In many instances, while this tree has been useful to explain the logic and depth of understanding needed to determine CCPs, it is not specific to all food operations, e.g., slaughter, and therefore it should be used in conjunction with professional judgement, and modified in some cases.

8. Establish critical limits for each CCP - (SEE PRINCIPLE 3)

Critical limits must be specified and validated for each Critical Control Point. In some cases more than one critical limit will be elaborated at a particular step. Criteria often used include measurements of temperature, time, moisture level, pH, Aw, available chlorine, and sensory parameters such as visual appearance and texture.

Where HACCP guidance developed by experts has been used to establish the critical limits, care should be taken to ensure that these limits fully apply to the specific operation, product or groups of products under consideration. These critical limits should be measurable.

9. Establish a monitoring system for each CCP - (SEE PRINCIPLE 4)

Monitoring is the scheduled measurement or observation of a CCP relative to its critical limits. The monitoring procedures must be able to detect loss of control at the CCP. Further, monitoring should ideally provide this information in time to make adjustments to ensure control of the process to prevent violating the critical limits. Where possible, process adjustments should be made when monitoring results indicate a trend towards loss of control at a CCP. The adjustments should be taken before a deviation occurs. Data derived from monitoring must be evaluated by a designated person with knowledge and authority to carry out corrective actions when indicated. If monitoring is not continuous, then the amount or frequency of monitoring must be sufficient to guarantee the CCP is in control. Most monitoring procedures for CCPs will need to be done rapidly because they relate to on-line processes and there will not be time for lengthy analytical testing. Physical and chemical measurements are often preferred to microbiological testing because they may be done rapidly and can often indicate the microbiological control of the product. All records and documents associated with monitoring CCPs must be signed by the person(s) doing the monitoring and by a responsible reviewing official(s) of the company.

10. Establish corrective actions - (SEE PRINCIPLE 5)

Specific corrective actions must be developed for each CCP in the HACCP system in order to deal with deviations when they occur.

The actions must ensure that the CCP has been brought under control. Actions taken must also include proper disposition of the affected product. Deviation and product disposition procedures must be documented in the HACCP record keeping.

11. Establish verification procedures - (SEE PRINCIPLE 6)
Establish procedures for verification. Verification and auditing methods, procedures and tests, including random sampling and analysis, can be used to determine if the HACCP system is working correctly. The frequency of verification should be sufficient to confirm that the HACCP system is working effectively. Verification should be carried out by someone other than the person who is responsible for performing the monitoring and corrective actions. Where certain verification activities cannot be performed in house, verification should be performed on behalf of the business by external experts or qualified third parties. Examples of verification activities include:
- Review of the HACCP system and plan and its records;
- Review of deviations and product dispositions;
- Confirmation that CCPs are kept under control.

Where possible, validation activities should include actions to confirm the efficacy of all elements of the HACCP system.

12. Establish Documentation and Record Keeping - (SEE PRINCIPLE 7)
Efficient and accurate record keeping is essential to the application of a HACCP system. HACCP procedures should be documented. Documentation and record keeping should be appropriate to the nature and size of the operation and sufficient to assist the business to verify that the HACCP controls are in place and being maintained. Expertly developed HACCP guidance materials (e.g. sector-specific HACCP guides) may be utilised as part of the documentation, provided that those materials reflect the specific food operations of the business. Documentation examples are:
- Hazard analysis;
- CCP determination;
- Critical limit determination.

Record examples are:
- CCP monitoring activities;
- Deviations and associated corrective actions;
- Verification procedures performed;
- Modifications to the HACCP plan;

An example of a HACCP worksheet for the development of a HACCP plan is attached as Diagram 3.
A simple record-keeping system can be effective and easily communicated to employees. It may be integrated into existing operations and may use existing paperwork, such as delivery invoices and checklists to record, for example, product temperatures.

TRAINING
Training of personnel in industry, government and academia in HACCP principles and applications and increasing awareness of consumers are essential elements for the effective implementation of HACCP. As an aid in developing specific training to support a HACCP plan, working instructions and procedures should be developed which define the tasks of the operating personnel to be stationed at each Critical Control Point. Cooperation between primary producer, industry, trade groups, consumer organisations, and responsible authorities is of vital important. Opportunities should be provided for the joint training of industry and control authorities to encourage and maintain a continuous dialogue and create a climate of understanding in the practical application of HACCP.

DIAGRAM 1
LOGIC CONSEQUENCE FOR APPLICATION OF HACCP

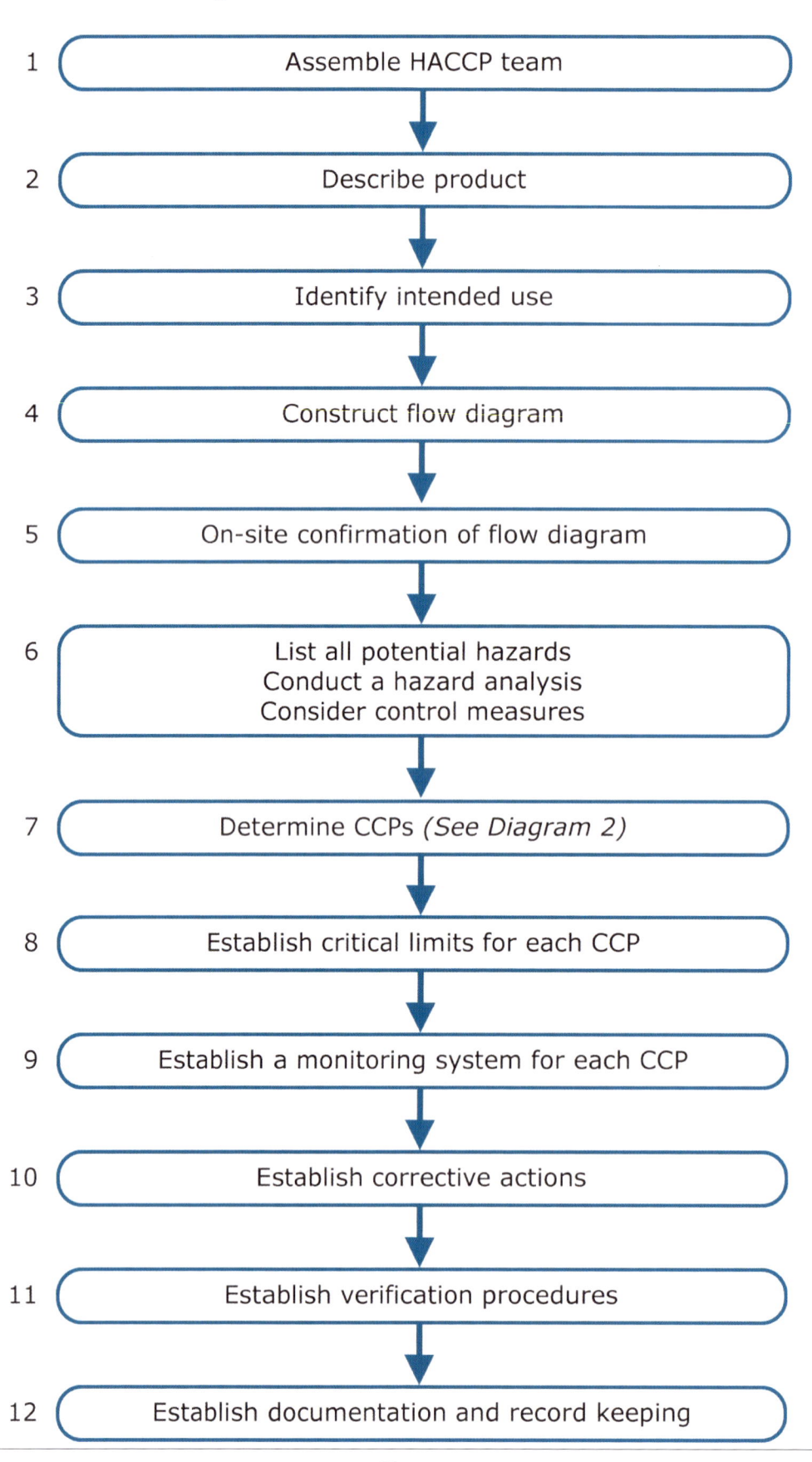

1. Assemble HACCP team
2. Describe product
3. Identify intended use
4. Construct flow diagram
5. On-site confirmation of flow diagram
6. List all potential hazards
   Conduct a hazard analysis
   Consider control measures
7. Determine CCPs *(See Diagram 2)*
8. Establish critical limits for each CCP
9. Establish a monitoring system for each CCP
10. Establish corrective actions
11. Establish verification procedures
12. Establish documentation and record keeping

DIAGRAM 2
EXAMPLE OF DECISION TREE TO IDENTIFY CCPs
(answer questions in sequence)

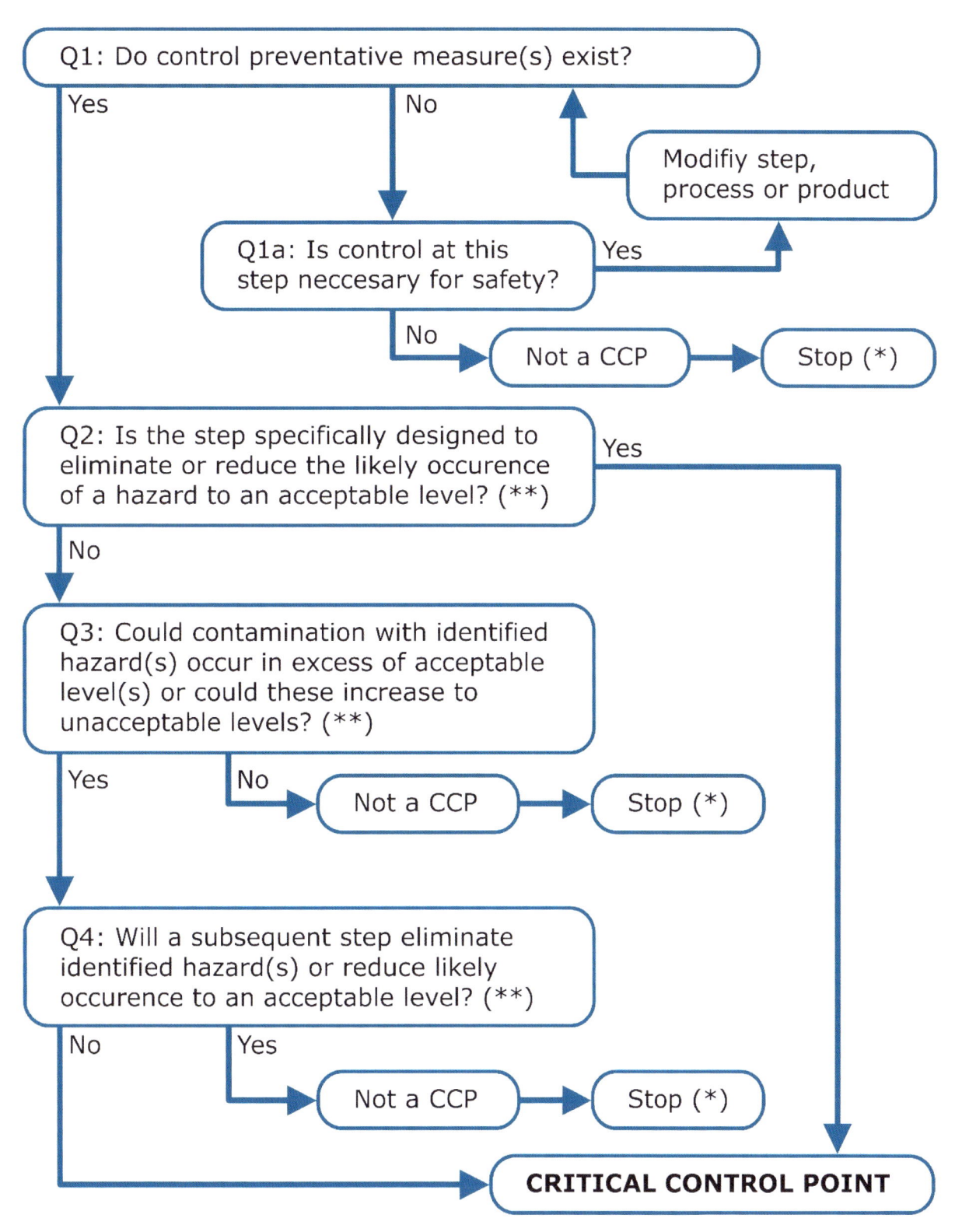

Q1: Do control preventative measure(s) exist?

- Yes → (go to Q2)
- No → **Q1a:** Is control at this step neccesary for safety?
  - Yes → Modifiy step, process or product → (back to Q1)
  - No → Not a CCP → Stop (*)

Q2: Is the step specifically designed to eliminate or reduce the likely occurence of a hazard to an acceptable level? (**)

- Yes → CRITICAL CONTROL POINT
- No → (go to Q3)

Q3: Could contamination with identified hazard(s) occur in excess of acceptable level(s) or could these increase to unacceptable levels? (**)

- Yes → (go to Q4)
- No → Not a CCP → Stop (*)

Q4: Will a subsequent step eliminate identified hazard(s) or reduce likely occurence to an acceptable level? (**)

- No → CRITICAL CONTROL POINT
- Yes → Not a CCP → Stop (*)

(*) Proceed to the enxt identified hazard in the described process.
(**) Acceptable and unacceptable levels need to be defined within the overall objectives in identifying the CCP of HACCP plan.

DIAGRAM 3
EXAMPLE OF A HACCP WORKSHEET

1. Describe product

2. Diagram process flow

3.

List							
Step	Hazard(s)	Control measure(s)	CCPs	Critical limit(s)	Monitoring procedure(s)	Corrective action(s)	Record(s)

4. Verification

Literatuur

British Retail Consortium, *BRC Wereldstandaard voor Voedselveiligheid*, versie 5. TSO (The Stationery Office), 2008.

Broek, M.J.M van den, Productvreemd gevaar langs de meetlat. Voedingsmiddelentechnologie (VMT) nr. 3, 2000.

Centraal College van Deskundigen HACCP, *Eisen voor een op HACCP gebaseerd voedselveiligheidssysteem*, vierde versie, Gorinchem: Stichting Certificatie Voedselveiligheid, 2006.

Codex Alimentarius, *Recommended International Code of Practice General Principles of Food Hygiene* CAC/RCP 1-1969, Rev. 4, 2003.

Codex Alimentarius, *Principles and guidelines for the conduct of microbiological risk assessment*, CAC/GL-30, 1999.

Codex Alimentarius, Principles for the establishment and application of microbiological criteria for foods. CAC/GL 21, 1997.

Cox, L.A., 'What wrong with risk matrices', Risk Analysis, Vol. 28, No. 2, 2008 'p.' 497-511.

Haagsma, J.A., Zanden, B.P. van der, Tariq, Pelt, L. W. van, Duynhoven, Y.T.P.H. van, Havelaar, A.H.,*Disease burden and costs of selected foodborne pathogens in the Netherlands, 2006 (Ziektelast en kosten van geselecteerde voedsel-overdraagbare micro-organismen in Nederland, 2006)*, Bilthoven: Rijksinstituut voor Volksgezondheid en Milieu, 2009. RIVM rapport 330331001.

Hekman, R., *Best Practice voor risicobepaling in een HACCP-plan*, Wageningen: Wageningen Universiteit, 2002.

ICMSF (International Commission on Microbiological Specifications for Foods), *Microorganisms in Foods 3: Microbial Ecology of Foods brings food spoilage and health risks into sharper focus through its study of how different food processes, ingredients, and product characteristics affect the microflora of foods. Vol. 1: Factors affecting life and death of microorganisms*, New York: Academic Press, 1980.

ICMSF (International Commission on Microbiological Specifications for Foods), *Microorganisms in Foods 4: Application of the Hazard Analysis Critical Control Point (HACCP) System to Ensure Microbiological Safety and Quality*, Oxford: Blackwell Scientific Publications, 1988.

ICMSF (International Commission on Microbiological Specifications for Foods), *Microorganisms in Foods 6: Microbial Ecology of Food Commodities*, New York: Kluwer Academic & Plenum Publishers, 2005.

International Featured Standards, IFS Food, version 4. 2007.

International Organization for Standardization , ISO 22000:2005, Nederlandse norm Voedselveiligheid managementsystemen , Eisen aan een organisatie in de voedselketen, Delft: Nederlands Normalisatie-instituut, 2005.

Kreijl, C.F. van,. Knaap, A.G.A.C. (red.), *Ons eten gemeten, Gezonde voeding en veilig voedsel in Nederland*, Bilthoven: Rijksinstituut voor Volksgezondheid en Milieu, 2004. RIVM-rapportnummer: 270555007.

Kujawska, O., *Bow Tie, an elegant solution for food processing companies. How the Bow Tie method can be used to get grip on hazards in the food industry*, Dronten: CAH Proffesional Agricultural University, 2008.

Notermans, S., *Dossier Voedselveiligheid, editie 2009*, Zeist: Stichting Food Micro & Innovation, 2009.

Printed in Great Britain
by Amazon